Electron Acceleration
in the Aurora and Beyond

Cover illustration

Adaptation of all-sky photograph of an auroral arc with a huge fold, taken from Andøya in northern Norway. The zenith is approximately top centre of the central panel, with the southern horizon at the bottom. The corner panels show the arc extending to the eastern horizon (original photgraph by T Edwards).

Electron Acceleration
in the Aurora and Beyond

D A Bryant

Rutherford Appleton Laboratory

INSTITUTE OF PHYSICS PUBLISHING
BRISTOL AND PHILADELPHIA

© IOP Publishing Ltd 1999

British Library Cataloguing-in-Publication Data

A catalogue record for this book is available from the British Library.

ISBN 0 7503 0533 9

Library of Congress Cataloging-in-Publication Data are available

Published by Institute of Physics Publishing, wholly owned by The Institute of Physics, London

Institute of Physics Publishing, Dirac House, Temple Back, Bristol BS1 6BE, UK

US Office: Institute of Physics Publishing, The Public Ledger Building, Suite 1035, 150 South Independence Mall West, Philadelphia, PA 19106, USA

Typeset in TEX using the IOP Bookmaker Macros
Printed in the UK by J W Arrowsmith Ltd, Bristol

To Norma

Contents

Preface

I should like to share with the reader some thoughts on how the electrons found in the high atmosphere, and in the space around our planet and beyond, came to acquire their speeds and energies. The views contained here are personal ones, born of some 20 years spent experimenting with and puzzling over these matters. The arena to be explored is vast. Acceleration sites include the aurora borealis, the Earth's magnetosphere and radiation belts, the solar wind, the Sun and the Cosmos. Electron energies extend from the lowest currently measurable to the highest conceivable. Mystery prevails at every stage.

The subject of acceleration in space is not yet a mature one, and neither, do I claim, is my understanding of it. Many of the ideas and conclusions are presented for the first time here, so this is not a textbook in the usual sense, but rather a discussion document. Some of the views run contrary to received opinion, but, I believe, always remain true to established physical principles. Emphasis throughout is on how these principles govern the varied phenomena on display. Many parallels will be found linking acceleration in space to particle accelerators in the nuclear-physics laboratory and even on the sports field. The approach I have adopted is predominantly illustrative and graphical, though I encourage direct involvement by the reader through quantitative formulae and some numerical and statistical experiments.

If, in these pages, I help in any way to encourage readers to draw their own conclusions about the acceleration of electrons and other charged particles in space, I will consider my aim to have been achieved.

Duncan Bryant
September 1998

Acknowledgments

The ideas expressed in these pages, for which I accept full responsibility, would nevertheless not have come into being without the imagination, skill and dedication of the many colleagues who have accompanied me or permitted me to join them in a variety of experiments and theoretical studies made over a substantial number of years.

A great debt is owed to my fellow experimenters in a number of rocket and satellite studies of charged particles of the aurora and other regions of space—Martin Courtier, Trefor Edwards, David Hall, Alan Johnstone, Chunky Lepine and Dick Mason—for their major contributions to the sound foundation of trustworthy measurements and careful interpretation upon which the ideas expressed here could be built. The valued collaboration of my fellow originators of the wave theory of the aurora, Robert Bingham and David Hall, is gratefully acknowledged. I am especially indebted to Chris Perry for invaluable discussions and assistance during the preparation of the manuscript. I thank Gerhard Haerendel and Tom Krimigis for the invitation to participate in the AMPTE mission, and my colleagues at the Rutherford Appleton Laboratory and elsewhere throughout the United Kingdom who designed, produced and operated a third satellite for this unique mission. The participation of the many co-authors of publications to which reference is made has been most influential and is greatly appreciated.

I thank the CLRC Rutherford Appleton Laboratory for the support and facilities which enabled the research programmes fuelling these pages to be carried out.

This work could not have been completed without the substantial help and constant encouragement of my wife, Norma.

Introduction

Electrons that have managed by some means to acquire energies far in excess of those of their fellows constantly flaunt their good fortune in perplexing displays of the northern and southern lights (the aurora borealis and aurora australis). The power brought from above into the upper atmosphere in these events, which are commonplace in the polar regions, is far greater than anything that could be managed artificially, being the equivalent of some ten or more large power stations. It would be only a slight exaggeration to say that we have not the faintest idea how this comes about. Cosmic rays, in the form of atomic nuclei and electrons, arrive with a steady flux at the Earth from unknown sources, having been accelerated far more than any of the world's most powerful accelerating machines could ever hope to emulate. It is no exaggeration at all to say in this case that we have not the slightest idea how Nature accomplishes this astonishing feat. The Sun taunts us with similar puzzles. At unpredictable and irregular intervals the Sun's hot outer atmosphere, or corona, erupts in what seem to be isolated explosions or electrical discharges known as solar flares. These flares throw freshly accelerated atomic nuclei and electrons into interplanetary space. Some of these find their way into the Earth's upper atmosphere where they upset the balance of the ionised layers and, as a consequence, disrupt radio communications. No-one knows how solar flares accelerate these particles. In fact, charged-particle acceleration by natural forces remains very much a mystery wherever and whenever it occurs.

Why is acceleration in space so hard to understand? After all we know how to build giant accelerating machines in the laboratory. These are used to hurl particles at one another at very high speeds in order to break them apart to reveal in the debris what the particles are made of. The dentist's x-ray machine works by accelerating electrons into a target which causes them to ricochet, radiating x-rays as they go. Our television sets work in a not too dissimilar way, but here energies are lower and the target screen fluoresces following the energy gained from acceleration being released on impact. We shall try to find some explanation for the continuing mystery of acceleration in the natural world in the following pages and, in so doing, try to gain an insight into what might be going on. For the moment we may note that in Nature we are dealing not with single particles of well defined energy, but in the main with assemblies, or distributions, of particles and, moreover, that it is rarely, if ever, possible to follow the changes that occur between the particles' source and the point where they are observed.

Chapter 1 provides the tools and concepts. We recall the basic properties of the electron and define exactly what is meant by acceleration, both in the general sense and in the sense of increased kinetic energy. The fundamental requirements of an accelerator are then derived. The concept of an assembly, or distribution, of particles will be explored. We shall consider how particular distributions might be generated and modified. The power of Liouville's theorem of continuity is witnessed in several manifestations. The properties—some quite unexpected—of purely random processes are looked into, together with the concept of entropy. Experimentation by the reader is encouraged with the inclusion of QBasic programs in Appendices 1 and 2. We summarise the methods used to sample electron distributions and comment on the relative merits and interpretation of various forms of display.

In Chapter 2 we survey the arena to be explored. We begin by describing the state of matter to which it belongs—the collisionless plasma, whose constituent charged particles interact predominantly through electric and magnetic forces. The concept of magnetic lines of force—to feature strongly in what follows—is then discussed. We then survey briefly the particles and fields of the solar wind and interplanetary space, and consider the interaction of the solar wind with planets and other bodies in its path. The nature of and processes governing the enclaves, or magnetospheres, formed as a result are then discussed. This is followed by a description of the Earth's magnetosphere and its electron content.

Our acceleration studies begin in Chapter 3 with the aurora. Here, just on the threshold of outer space, the facts are particularly well documented. Following a description of the aurora and the many and varied facts known about the electron streams responsible, we set out to discover the underlying processes. We look into the currently widely held explanation for the acceleration of electrons responsible for the brightest types of aurora. We soon notice that this 'explanation'—the potential difference theory—fails to satisfy the fundamental requirements for an accelerator. An attempt is made to trace the events leading to such a curious state of affairs and evidence for its invoked electric fields is sought. A different interpretation—the wave theory—based on Landau's far-reaching ideas on wave–particle interactions, and developed by my colleagues and myself following our realisation of these problems some decades ago, is then discussed and evidence for its invoked electrostatic waves is sought. Exploration of the key properties of resonant interactions between charged particles and waves is encouraged here too with a QBasic program given in Appendix 3.

Next, we move out into the solar wind for three studies of electron acceleration caused by local perturbation of its flow. The first, in Chapter 4, centres on the shock wave—the bow shock—upstream from the Earth's magnetosphere, the second, in Chapter 5, on obstacles created artificially by the release of clouds of lithium and barium ions and the third, in Chapter 6, on a somewhat mysterious natural version of the ion experiments. Theories put forward to account for the electron acceleration at these sites are discussed within each chapter, where it is again evident that some are fundamentally lacking. A modified version of our auroral wave theory is found to yield promising preliminary results.

We move, in Chapter 7, to the boundary between the perturbed solar wind—the magnetosheath—and the Earth's magnetosphere, a boundary known as the magnetopause. A new technique, owing much to the solving of jigsaw puzzles, is employed to reveal the structure of the boundary layer of transition between the plasmas of the magnetosheath and magnetosphere. An interpretation, again in the form of acceleration by waves, is suggested to account for clear evidence of electron acceleration at this site also.

We enter, in Chapter 8, the Earth's magnetosphere to review the various electron populations, their prospective sources and acceleration mechanisms. This is the stage at which we begin to encounter electrons whose speed approaches that of light and where the effects of relativity need to be taken into account. The fluctuating electric and magnetic fields of and within the magnetosphere will be seen to make very capable particle accelerators. Theories invoking accelerators that do not meet fundamental requirements lurk here too.

The understanding of the acceleration processes at the Sun, discussed in Chapter 9, which produce the solar wind and solar-flare electrons is made very difficult by the fact that we have no knowledge of these particles before they were accelerated. Nevertheless, some tentative conclusions can be drawn. Here, we meet for the first time a new adaptation of one of the most famous of all natural acceleration mechanisms—Fermi's mechanism for the acceleration of cosmic rays. With cosmic rays, in Chapter 10, we encounter the most intractable of all acceleration problems. Not only do we lack information about the source distribution, we have only scant perception of the medium through which the accelerated particles travel for astronomical times before reaching us. Even then the fluxes are so low as to be barely measurable. Nevertheless, application of Fermi's ideas and the basic principles of Chapter 1 help us to gain some insight into what may be going on, and at the same time develop a further understanding of the surprising nature of random processes in this and other contexts.

The many threads are drawn together in Chapter 11, with inevitably some frayed ends betraying our present limited understanding. Some general questions are posed and exercises suggested at the end of the work. Answers are suggested.

The scheme I have adopted to keep physics to the fore is descriptive and graphical. The text avoids all but the bare minimum of mathematical manipulation. The quantitative aspect is provided by formulae tailored to the present application. These and other related formulae are collected together in Appendix 4.

The international standard SI metre–kilogram–second system of units is employed throughout with convenient derivatives. In order to avoid unnecessary clutter in this application, where statistical fluctuations are rarely an issue, error

Chapter 1

Acceleration principles

This opening chapter sets out the basic properties of the electron and the physical principles that will guide the explorations to follow.

1.1 The electron

The electron is one of the fundamental particles of matter. It is the lightest fundamental particle endowed with an electric charge.

1.1.1 Basic properties [1]

The electron has a mass

$$m_e = 9.11 \times 10^{-31} \text{ kg}$$

and carries an electric charge of e, where

$$e = -1.60 \times 10^{-19} \text{ C}.$$

The charge-to-mass ratio, of

$$e/m_e = 1.76 \times 10^{11} \text{ C kg}^{-1}$$

is the highest of any particle—a distinction the electron shares with its anti-matter mirror image, the positron. Since response to electric and magnetic fields increases with this ratio, just as it does for the sail-to-mass ratio for sailing craft, the electron could be considered to be the racing yacht of the charged particle world. Herein lies the liveliness of these particles and our particular interest in them. Electrons can exist bound by electrostatic attraction to atomic nuclei, they may be relatively free to move within electric conductors, or completely free to roam among and be buffeted by the electric and magnetic fields of their neighbourhood.

By virtue of its mass an electron at rest has a rest energy of

$$E_0 = m_e c^2$$
$$= 8.18 \times 10^{-14} \text{ J}$$
$$= 5.11 \times 10^5 \text{ eV}$$

c being the speed of light *in vacuo*, 3.00×10^8 m s^{-1}.

1.1.2   Attributes of motion

An electron with a velocity magnitude, or speed v, has kinetic energy, the quantity at the centre of our discussion:

$$E = (\Gamma - 1)E_0$$

where

$$\Gamma = (1 - (v/c)^2)^{-1/2}$$

is a factor which takes into account the effects of relativity which become important close to the speed of light.

Electron kinetic energies over the full range we shall be considering in our exploration, 1 eV to 10^{12} eV, are plotted in figure 1.1 against the corresponding speeds from 5.93×10^5 m s^{-1} to 3.00×10^8 m s^{-1}. The plot is seen to fall into two main sections.

In the non-relativistic (NR) range, well below the speed of light, E increases as the square of the speed;

$$E_{NR} = \tfrac{1}{2}m_e v^2.$$

In the ultra-relativistic (UR), close to the speed of light, $E \gg E_0$;

$$E_{UR} = \Gamma E_0.$$

An electron in motion has a momentum

$$p = \Gamma m_e v.$$

Momentum may be expressed in terms of E and E_0, both in eV, as

$$p = 5.34 \times 10^{-7}\sqrt{E^2 + 2E E_0} \text{ zN s}$$

from which it follows that

$$p_{NR} = 5.4 \times 10^{-4}\sqrt{E} \text{ zN s}$$

and

$$p_{UR} = 5.34 \times 10^{-7}E \text{ zN s}.$$

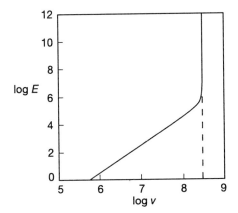

Figure 1.1. Relation between electron speed and kinetic energy. The log of kinetic energy in eV is shown against the log of speed in m s^{-1}. At low speeds energy increases as the square of the speed, while at high, relativistic speeds energy increases more and more rapidly as the limiting speed of light (dashed line) is approached.

1.2 Acceleration

Acceleration in its most general sense means the rate of change of velocity, i.e.

$$a = dv/dt.$$

Since velocity involves both a speed (the magnitude of the velocity) and a direction, the term 'acceleration' applies in general to a change in either or both of these properties. Because of the nature of the questions we shall be trying to answer, the 'acceleration' of our title refers, though, to the specific case of a gain of speed, i.e. an increase in kinetic energy. Since speed, like velocity, is a relative quantity, speed and kinetic energy will generally be seen differently by different observers.

Newton's laws inform us that to induce or change motion we need to apply a force, $\mathcal{F}$. The acceleration is then

$$a = \mathcal{F}/m$$

in the direction of the applied force. The quantity of kinetic energy dE transmitted to the accelerated object in each element of path dl is the product of the path length and the component of force along the direction of motion, a direction which may well change during the interaction. So

$$dE = ma\,dl\cos\theta$$

where θ is the angle between the force and the instantaneous direction of motion.

Kinetic energy, once gained, remains an attribute until it is stored, usually temporarily, as potential energy, or transferred to other objects. Any component of force perpendicular to the motion changes the direction (and thereby the velocity) but not the speed and so does not change kinetic energy.

Reduction in speed also changes kinetic energy, and a gain by one object may well be dependent upon retardation of another. We shall, however, because of the nature of this investigation, be less concerned here with the many ways electrons and other particles can lose kinetic energy than with the ways they become accelerated to higher speeds. We shall be looking, then, for processes that increase the kinetic energy of electrons.

1.2.1   What is an accelerator?

We shall now try to establish what makes an accelerator, and to find the effects of different types of accelerator on individual charged particles.

A particle accelerator, in the terms we are discussing, is an agent capable of delivering energy. Transfer of a given quantity of energy can be accomplished by a force being applied over the requisite distance, or, equivalently, being applied for the necessary time. The force must have a component along the particle's direction of motion, a direction which may change during acceleration. There is also another requirement. It is that the accelerator must perform a net quantity of work on the particle. An agent or mechanism that is able only to repay kinetic energy which it has previously taken away does not provide energy, any more than banks provide us with money. Similarly, one which takes away what it has earlier provided as does a depression encountered while cycling along an undulating road, cannot be classed as an accelerator either.

While these considerations might seem obvious and uncontroversial, as indeed they should be having been cornerstones of physics since the 17th Century, they do sometimes provoke a challenge, similar in some respects to those made in the search for perpetual motion. The discussion might go something like this ...

'But, what if an object falls into a hole or depression? Its kinetic energy there is unquestionably greater than it was before entry, and can be used to do work when it reaches the bottom.'

'Fine! Let us test this idea by converting the kinetic energy into electricity by driving a dynamo. All we would need, according to this scheme, is to let material, say water, pour down and drive the dynamo. Problem: the hole will fill with water!'

'We can pump it out.'

'Wouldn't that take at least the same quantity of energy that the hole had "provided" in the first place, even with a perfectly efficient pump?'

'Perhaps we should dig another hole ...'

!!!

1.2.2   Acceleration by kinematic collision

It will help crystallise some of the points made above and guide us in later work
if we examine a specific case of acceleration, the acceleration caused by one
object colliding with another. Let us see what happens when two solid spheres of
equal mass roll towards each other, as shown in frame **a** of figure 1.2. In order to
keep things simple and to be able to deal only with the translational motion of the
spheres we shall neglect the complication of the energy of rotation by letting both
spheres have most of their mass concentrated at their centres. We could equally
well consider two masses sliding on a frictionless surface, to which stones in the
sport of curling approximate. We shall let both objects have, initially, one unit
of velocity, but in opposite directions. When they collide they will slow down
as their energy is stored temporarily as the potential energy of compression. If
the spheres are perfectly elastic, i.e. if no energy goes into permanent distortion
or into heat or sound (all of which are, of course, unavoidable in practice), the
spheres will spring back into shape, returning the kinetic energy borrowed, and
the spheres will be accelerated away from the point of contact until they acquire
exactly the same speeds that they had before but with velocities reversed. Since
the speeds, i.e. the magnitudes of the velocities, are unchanged, and the kinetic
energies before and after the interaction are the same, the process is not one of
acceleration in our (increase-of-kinetic-energy) terms.

We can now look at exactly the same sequence of events from a different
vantage point. If we move along with the white sphere and continue to view
events from this frame of reference, frame **b** of figure 1.2, we will see the white
sphere as being initially stationary, and the black sphere moving towards it,
i.e. to the right, at two units of velocity. At the point of collision we will see
both flattened spheres moving to the right at one unit of velocity. When they
spring apart the black sphere will appear stationary and the white sphere will
be moving at two units of velocity to the right. The net result in this frame of
reference, then, is that the white sphere has been accelerated from zero to two
units of velocity, and the black sphere has been retarded by the same amount.
Conversely, if we viewed events from the initial frame of reference of the black
sphere (frame **c**) we would perceive the black sphere to be accelerated and the
white one brought to rest.

These considerations show that a process which may result in no exchange
of energy in one frame of reference may well result in energy exchange in
another frame.

The three different views of the interaction, and indeed those from any other
frame, are captured conveniently in figure 1.3. Here the ordinate is velocity, v,
and the abscissa is position, x, along the line of motion (we are assuming for

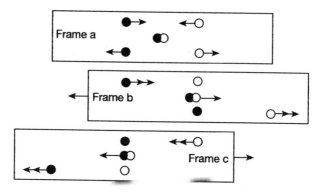

Figure 1.2. Three stages of an elastic collision viewed from three different frames of reference. The arrows indicate the magnitudes and directions of motion of black and white spheres of equal mass, and the frames from which their collision is viewed. Double arrow heads indicate doubled speed. In frame **a** the black and white spheres approach each other, come to rest momentarily at impact, and then separate at the same speeds that they had initially. There is no exchange of energy in this frame. In frame **b** the white sphere is at rest initially and gains kinetic energy from the collision. In frame **c** the black sphere gains energy.

simplicity that everything takes place along a straight line). The white sphere starts with velocity v_1, and the black one starts with the higher velocity v_2. At the point of collision both spheres have the same velocity—their 'centre-of-mass velocity' v_c—mid-way between v_1 and v_2 (since the spheres have equal mass). After the interaction the two objects, having interchanged velocities, leave the scene with the black sphere at v_1 and the white at v_2.

To view the event as it is seen from frame **a** of figure 1.2 we give ourselves a velocity v_c. From here we would observe the final speeds of both spheres to be the same as the initial ones, though with their directions reversed. This frame, in which the particles have equal and opposite velocities at all times is also the frame in which the sum of their momenta remains zero. This frame, which moves with the average position of the two masses, is known as the centre-of-mass frame.

We can simulate frame **b** of figure 1.2 by declaring v_1 to be zero, and measuring velocities as departures from v_1. Notice that all velocities are now either positive or zero, in accord with all arrows in frame **b** pointing to the right. The fact that the velocity of the white sphere increases throughout the event demonstrates that it has undergone a net acceleration. Frame **c** of figure 1.2 can be represented by defining v_2 to be the zero. All velocities are now either negative or zero in accord with the left-facing arrows of frame **c**. From this viewpoint, the white sphere is slowed down throughout. The view from v_1 is just that seen at a curling match as one stone slides into another.

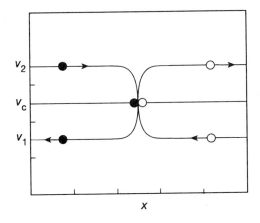

Figure 1.3. Generalised view of the interaction of figure 1.2. Speed v is plotted against position x. The view from frame **a** is obtained by measuring all velocities from v_c, the centre-of-mass velocity (see text). The view from frame **b** is obtained by measuring all velocities from v_1, and the view from frame **c** by measuring from v_2. Other reference frames may be chosen at will.

Shuffleboard collisions are also of this type. The principles at work here can all be demonstrated very effectively, too, with the bored-executive's multiple pendulum.

We can easily modify figure 1.3 to cope with cases where one sphere is heavier than the other. All we have to do is to find the velocity, v_c, which, taken as a reference, makes the momentum of the heavier (black) sphere equal and opposite to that of the lighter (white) sphere. In figure 1.4, where this is done, we view proceedings from this v_c, energy conserving, or 'conservative' frame, where initial and final velocities for both spheres, though reversed by the interaction, are unchanged in magnitude. In all other frames (if we restrict motion to be along the x direction) there is an exchange of energy, since the initial and final velocities are then different for both spheres.

For a non-relativistic particle of mass m and velocity v being reflected by a much-more-massive particle, or other form of reflector, moving towards it at velocity w, the energy gain is

$$\delta E_{NR} = 2mw(v + w).$$

Tennis provides examples of these asymmetric interactions. A serve has the spectator at velocity v_1 witnessing a stationary ball (white sphere) being struck by a moving racket (black sphere). A forehand or backhand drive has a spectator at v_2 seeing the ball reverse direction on contact. A drop shot appears to a spectator (and player) at v_3 as a reversal of direction with reduced speed. Spectators at v_1 might alternatively be observing the striking of a golf ball, a cue ball in snooker or pool, a nail by a hammer or a post by a pile driver. From

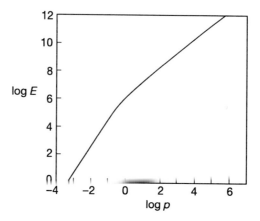

Figure 1.4. Generalisation of figure 1.2 for spheres of different mass, the black being the heavier. With a reference velocity of v_1, the figure illustrates the basic dynamics of a tennis serve, the black sphere representing the racket and the white sphere the ball. With reference velocity of v_2 we see a drive, and with v_3, a drop shot. With the reference v_4, from which all velocities, before and after the interaction, are in the same sense, we witness a sweep shot in cricket or the kicking of a football, along the initial direction of motion. These last, resonant, interactions are related to the principle on which surfing is based.

v_4 we could be observing a 'knock-on' event such as a sweep shot in cricket, or the kicking of a football along the direction in which it already moving. This view from v_4, which will play a major role in later discussion, also covers a spectator's view of surfing or rather what surfing would be like without the viscous losses that help to keep a surfer in touch with a wave.

The details of the interaction are immaterial. In fact the 'collision' could equally well be effected by mutual repulsion between two objects carrying the same sign of electric charge or of sharing the same magnetic polarity. Applications will be found in the studies to follow of electron acceleration in the aurora and in the solar wind.

The general expression, holding for any energy, obtained by an analysis in terms of momentum rather than velocity, is

$$\delta E = 2\beta \left(\sqrt{2EE_0} + \beta E_0 \right) \sqrt{1 + E/2E_0}$$

where $\beta \equiv w/c$. We shall have recourse to this formula when considering solar-flare electrons in Chapter 9.

At ultra-relativistic energies,

$$\delta E_{\text{UR}} = 2\beta E.$$

This last result forms an important part of Fermi's theory of cosmic-ray acceleration [2] to be discussed in Chapter 10.

1.2.3 Acceleration by electric fields

1.2.3.1 Static and unchanging fields

An electric field $\mathcal{E}$ imparts to a particle of charge-to-mass ratio e/m an acceleration

$$a = e\mathcal{E}/m.$$

In line with the above discussion for a general force, the energy imparted to a particle carrying a charge e over an element of path $\mathrm{d}l$ inclined at an angle θ to the field is

$$\mathrm{d}E = e\mathcal{E}\,\mathrm{d}l\cos\theta.$$

This is conveniently written in terms of the concept of electrostatic potential difference $\Delta\Phi$ as

$$\Delta E = e\Delta\Phi.$$

With this in mind let us consider the motion of a charged particle in the electric field produced by an assembly of stationary charged particles forming a region of space charge. The actual path followed by the particle will depend on a number of factors, including the initial motion, number and deployment of positive and negative particles in the assembly. However, there are some general points that can be made whatever the details. These follow from the fact that it would be possible, if we were told exactly where all the charges were, to assign to each and every point inside and outside the region an electric potential, which is the electrical equivalent of altitude in a gravitational field. Furthermore we know that all the points having the same potential would form one or more sets of closed surfaces. If we could see these equipotential surfaces, they would look like sets of balloons within balloons. Figure 1.5 illustrates this in two dimensions in which closed surfaces appear as closed loops.

One aspect of a field that can be so portrayed is that, if a charged particle starts a journey from any point, say **A** in figure 1.5, and returns to the same point, it returns with exactly the same kinetic energy it had when it left, because $\Delta\Phi = 0$. This is the case whatever the route taken and however the velocity changes *en route*. The effect is equivalent to the gravitational analogy of throwing a ball into the air and finding that, apart from the inevitable slowing down by air resistance, it returns with the same speed.

Another consequence of the conservative nature of unchanging electric fields is that if a charged particle returns to another place where the potential is the same, e.g. **B** to **C**, it again arrives, despite temporary changes, with its initial kinetic energy. Within the present analogy, this is the case of catching a ball thrown by someone standing at the same level.

There is still a third consequence, which is really a special case of the second. Since the attraction or repulsion of electric charges diminishes as distance increases, it follows that the ups and downs of electric potential become weaker and weaker as we move further and further from the assembled charge.

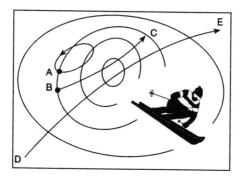

Figure 1.5. Illustrating the conservative nature of a potential field such as an electrostatic or gravitational one. The closed curves are 2d cross sections of the closed surfaces of common, or equi-potential. Journeys (arrowed) such as the round trip from **A**, the traversal between **B** and **C** at the same potential and that between the remote points **D** and **E**, lead to no net change in kinetic energy. Whether a transition between different potentials, such as that of the downhill skier, provides acceleration is an issue to be discussed in relation to the 'established' interpretation of the acceleration of auroral and other electrons.

At distances that are so great that the region of space charge, when seen from there, fills only a small part of the field of view, the potential due to the space charge becomes negligible. This is the same in any and all directions, so a charged particle starting a trajectory at one remote point, **D**, and finishing at any other remote point, **E**, starts and finishes its journey at the same potential. The third consequence is, then, that encounters with regions of space charge made during journeys between points that are remote from a space-charge region do not change a particle's kinetic energy.

Now let us consider journeys from one potential to another, as in the analogy of a ball being thrown from the top of a building, or a downhill skier moving across gravitational equipotentials. Although kinetic energy is gained on both of these one-way journeys, it is clear from our earlier discussion that this is only a part of the story. The energy has to be paid for—in the case of the ball-throwing by a climb to the top of the building, and in the case of the skier by the ski lift or a climb of the mountainside. Perpetually recycling waterfalls and the like remain the stuff of comic drawings. Regions of static and unchanging space charge do not, therefore, constitute particle accelerators.

There are many practical particle accelerators employing static electric fields, but, reassuringly, they all need help from non-electrostatic, non-conservative elements. In a cathode-ray tube, by means of which J J Thomson discovered the electron just over 100 years ago [4], and television tubes, where electrons are accelerated from a few electron volts to several thousand electron volts by the electric field between the cathode and anode, the essential non-conservative element is a battery or dynamo. Within a battery electrons are driven against the internal electric field at the expense of the battery's chemical energy. In a dynamo, it is the mechanical energy that drives the dynamo that achieves the same end. The early particle accelerating machine, the Van der Graaf generator [5], although often described as an electrostatic device, is no exception. The machine works by spraying charged particles, by means of corona discharge (similar to lightning before the main strike), onto a moving belt. The belt carries the charge to an insulated sphere to which the charge is transferred, again by corona discharge. The sphere becomes charged to a voltage which can be several million volts relative to its surroundings. Particles having the same sign of charge as the sphere and released from there are then accelerated back to ground potential where, having gained kinetic energies of several million electron volts for each unit of charge that they carry, they may be used as projectiles for nuclear interactions. In some applications they are used as a source for further acceleration by other types of accelerator. While it is undeniable that the Van der Graaf generator employs a static electric field to accelerate particles, it is clear that, just as in the case of the cathode-ray tube, the machine does not work by electrostatic forces alone. The drive belt is the equivalent of the battery or dynamo of the former case, since it also conveys charge against the electric field to set things up initially. Without the drive belt, the machine does not work.

Despite the fundamental conservative nature of electrostatic fields, and indeed all other fields that can be represented by sets of equipotential surfaces, there is, astonishingly, still an electrostatic configuration that many in the field of auroral and magnetospheric physics believe to be capable of accelerating charged particles. This is the electrostatic double layer [6, 7]. A double layer consists of two parallel discs of opposing electric charge separated by a distance that is small compared to the disc radius. There is, as a consequence, an electric field in the space between the disks able to accelerate charged particles across the gap. However, it will be clear from our earlier discussion that before charged particles can reach the disc from which they are to be accelerated, they must first overcome the repulsion experienced in getting there. Although the electric field outside the gap will be weaker than the field inside, it will be correspondingly more extensive, as the example of figure 1.6 shows. The external field will ensure that, in overcoming this repulsion and the restraining drag on the emerging side, charged particles will lose just as much kinetic energy as they gain on being accelerated across the gap. There is no escape from this. The double layer cannot accelerate anything. Figure 1.7 shows in a three-dimensional projection the spatial distribution of electric potential due to a double layer, showing that the potential at large distances is the same in all directions, and confirming that encounters with double layers, being just a particular case of finite regions of static space charge, will not result in acceleration. There is a clear connection between this figure and the earlier thoughts about cycling along an undulating road.

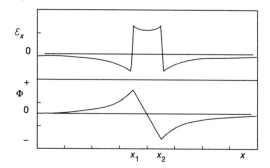

Figure 1.6. Electric field and potential in the region of a double layer consisting of a positively charged disc at x_1 and a negatively charged disc at x_2. The electric field plotted is the axial field and the potential is measured on the axis of symmetry. The electric field between the discs is stronger, but less extensive, than the oppositely directed field outside. The potential reaches a maximum at x_1 and a minimum at x_2. There is no net change of potential across the system.

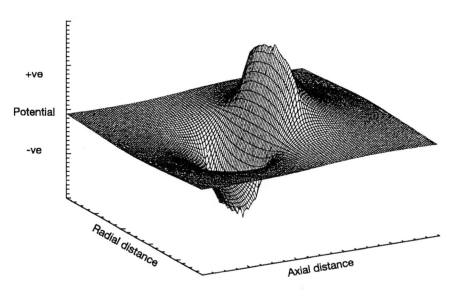

Figure 1.7. Three-dimensional representation of electrostatic potential in the region of a double layer. Note that the potential excursions are localised (courtesy C H Perry).

How is it, then, that a belief has arisen, as we shall discuss in Chapter 3, that double layers can serve as accelerators of electrons in the aurora and elsewhere? A possible explanation is that if a double layer is treated as a pair of discs of infinite radius, the internal electric field will be equal to the potential difference between the discs divided by the spacing, and the external field will be zero. Equipotentials in this picture will be flat sheets parallel to the discs, extending to infinity. This treatment overlooks the vital fact that, for any real double layer, which must be finite in size, the external electric field is not zero. It seems that Alfvén, the originator of the idea, was well aware of this problem when he defined a double layer as being the simplest configuration of electric charge that produces an electric field inside the layer and a zero field outside [8]. The snag is that there is always a field outside.

1.2.3.2 Static and unchanging fields in a moving frame

We can break out of the conservative impasse if we consider the interaction between charged particles and an assembly of space charge in motion.

To investigate, we examine a deflection of a charged particle caused by repulsion from a space-charge region made up of very many particles. For simplicity we consider the particle's motion and space-charge motion to be co-planar. In figure 1.8 we view proceedings from a frame of reference moving at the same velocity as the space-charge assembly. In this, conservative, frame the particle's trajectory is symmetric. The velocity parallel to the line of symmetry undergoes no net change, but is reversed in direction by the encounter. The velocity perpendicular to this is the same before and after deflection. In figure 1.9 we view the same event from a frame in which the space-charge region has a velocity w. Now, although there is again no net change in perpendicular velocity, the parallel velocity is not only reversed but increased in magnitude by twice the speed of the space-charge region.

It will be immediately apparent that the interactions depicted in figures 1.8 and 1.9 are dynamically equivalent to the earlier-considered collisions between heavy and light objects. This realisation is central to this work. It enshrines the basic physics of the interaction between charged particles and waves of electric potential, electrostatic waves or magnetised regions in motion. Loss of energy by waves to charged particles causes the waves to diminish or be damped down in a process named after its predictor as Landau damping [9]. Electrostatic waves of a particular type we shall discuss in some detail in later chapters—lower hybrid waves—are employed in the acceleration of electrons in the drive towards controlled nuclear fusion. The linear accelerator, or linac [10] (figure 1.10), is based on the same principle. Here, charged particles, or more strictly bursts or bunches of charged particles, are accelerated in a straight line through the device by a series of electrodes whose potential is varied at a fixed rate, but whose length increases with distance along the accelerator to ensure that charged particles passing along the axis repeatedly find themselves in an

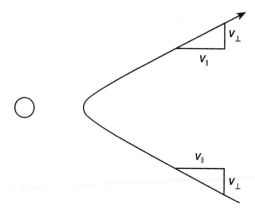

Figure 1.8. Deflection of a moving object by repulsion. The trajectory (arrowed) is seen here in the frame of reference of the centre of repulsion (white sphere). The net effect is for the velocity perpendicular to the line of symmetry to be unchanged and for the parallel velocity to become reversed. There is no net change in speed in this frame of reference.

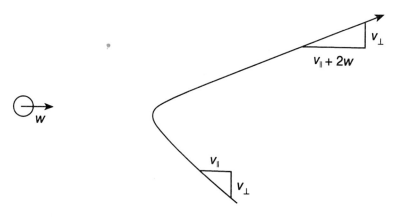

Figure 1.9. The deflection of figure 1.8 seen from a frame of reference in which the centre of repulsion is moving at velocity w. Here, the parallel velocity of the deflected object undergoes not only a reversal in direction but also a magnitude increase of $2w$.

accelerating electric field. This is an example of surfing on a coherent wave, and is closely related to the surfing on electrostatic wavepackets to feature strongly in later chapters.

For completeness we should mention another class of particle trajectory that may be allowed by a static or moving space-charge configuration. This is a trajectory of a charged particle that has become trapped in a hollow or potential well within a space-charge region. Notice that a particle cannot drift

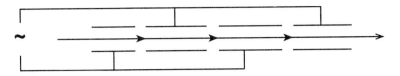

Figure 1.10. Principal of the linear particle accelerator, or linac. An oscillating voltage is used to propagate a wave of electric potential along a series of coaxial tubes. Charged particles travelling along the axis in phase with the wave are accelerated at each of the gaps (bold arrows). The drift paths between the gaps increase along the accelerator to maintain resonance between the wave and the accelerating particles.

into entrapment, since, if it has enough kinetic energy to find its way into a potential well, it also has enough to escape again. An example of this particular difficulty—in fact, an impossibility in purely conservative fields—is the need to fire rocket motors when a satellite is injected into orbit around the Earth or other planet. A more homely example is that of lodging ball bearings in the shallow depressions in games found in Christmas crackers. However, given that there is a potential well, it is always conceivable that particles may have become trapped within it when it was formed, or that they were created by ionisation within it, or, perhaps, that a third party may have perturbed the trajectory to establish trapping. It is clear, though, that, although a trapped particle's velocity will almost certainly oscillate about any velocity the potential well has, the mean velocity will remain at that of the potential well, and there will be no acceleration, unless the well itself accelerates.

The main conclusion to be gained from this discussion is that moving patterns of electric potential can promote acceleration.

1.2.3.3 Time-varying fields

Electric fields that vary with time cannot, except in special circumstances (see Chapter 3), usefully be treated as gradients of electrostatic potential. This is because electrostatic potential surfaces would need to be updated continuously, to keep pace with a changing pattern of space charge. The escape, thus afforded, from the straight-jacket of conservative fields offers, therefore, another possibility for particle acceleration. Wherever time dependent fields are found, acceleration, or retardation, will not be far away.

1.2.3.4 Feedback

We have so far been discussing electric fields and particles moving within them as though these were independent entities. In practice, the redistribution of charged particles caused by acceleration will modify the causative electric field. Moreover, the modification will always tend to reduce the field strength. A vital aspect, therefore, of any accelerator, whether natural or artificial, is the means of replenishment of its fields and source particles.

1.2.4 Acceleration by magnetic fields

1.2.4.1 Static

Acceleration by a magnetic field of flux density, or 'strength', B is given by

$$a = evB\sin\alpha/m$$

where α is the angle between the field direction and the instantaneous direction of motion of the particle, known as the pitch angle. Acceleration is perpendicular to both v and B.

It is often convenient to consider particle velocities and momenta as having components parallel $\parallel$ and perpendicular $\perp$ to B, i.e.

$$v_\parallel = v\cos\alpha \qquad v_\perp = v\sin\alpha$$
$$p_\parallel = p\cos\alpha \qquad p_\perp = p\sin\alpha.$$

The general motion of a charged particle in a uniform magnetic field is in the form of a helix composed of a gyration at constant $v_\perp$ and $p_\perp$ about the field direction, combined with a steady motion $v_\parallel$ and $p_\parallel$ along the field direction. The radius of gyration, gyro-radius, or cyclotron radius of a particle of Z units of electronic charge, kinetic energy E and rest energy E_0 moving with a pitch angle α through a magnetic field of B nT is

$$r_c = 3.33\sin\alpha\sqrt{E^2 + EE_0}/ZB \text{ m}$$
$$= 6.24 \times 10^{27} p_\perp/ZB \text{ m}.$$

The angular gyro-frequency ω_g is

$$\omega_c = eB/\Gamma m$$

which for non-relativistic speeds is independent of energy, but which varies inversely as kinetic energy at relativistic speeds.

For electrons in a magnetic field of B nT

$$\omega_{ce} = 1.76 \times 10^2 B/\Gamma \text{ rad s}^{-1}$$

and the gyro-period $\tau_{ce} = 2\pi/\omega_{ce}$, is

$$\tau_{ce} = 3.57 \times 10^{-2}\Gamma/B \text{ s}.$$

Since the acceleration is always perpendicular to the direction of motion, no energy is imparted. Static magnetic fields are therefore not particle accelerators. Motion in such fields is conservative just as it is in unchanging electrostatic fields.

1.2.4.2   Static in a moving frame

Magnetic fields in motion, even if the configuration remains fixed, can deliver kinetic energy to charged particles in an analogous way to a moving, but otherwise unchanging pattern of electric field. The dynamics are the same in both cases. Enrico Fermi's celebrated theory of cosmic-ray acceleration [2] is based on just this principle. In a process we shall explore more closely later, Fermi appealed to just this interaction taking place between 'wandering' interstellar magnetic fields and cosmic-ray particles encountering them. While the randomness of such collisions would ensure that both energy-gaining and energy-losing exchanges would occur, some particle could be expected to acquire very high energies through being involved in more energy-gaining interactions than energy-losing ones because approaching encounters are statistically more likely than overtaking ones. We shall call upon this and other aspects of this important mechanism in several different contexts.

1.2.4.3   Time variation

A magnetic field that changes with time gives rise to an electric field [3], and thereby to an acceleration

$$a = \frac{e}{m} \, dB/dt.$$

Acceleration is perpendicular to the magnetic field direction and will in general have a component along the particle's path, thus doing work on the particle and changing its kinetic energy. The kinetic energy increment dE along an element of path dl is given by

$$dE = e \, dl \cos\theta \, dB/dt$$

where θ is the angle between dl and the induced electric field.

Energy will, in general, be gained (or lost) even when a particle executes a round trip through a medium where the field is changing; the gain of energy being

$$\delta E = e \, d\phi/dt$$

where ϕ is the magnetic flux enclosed, uniform or otherwise. The concept of potential, so valuable for electrostatic fields, is not helpful for induced electric fields. The crucial difference between the electric fields from the two sources lies not in the fields themselves but in the configurations they adopt. Figure 1.11 shows that the electric field surrounding a region of changing magnetic field takes the form of concentric circles, and that electrons passing the region come to experience a net accelerating or retarding force, depending on the route they take.

Changing magnetic fields are employed in the particle accelerator known as the betatron [11]. The magnetic field serves two purposes in this device. One is to guide particles around closed (or quasi-closed) paths, while the other is to induce an accelerating electric field (we shall discuss this further

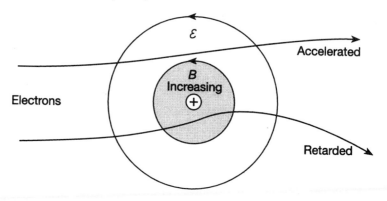

Figure 1.11. Motion of electrons in a region where the magnetic field increases with time. The changing magnetic field in the dark central region induces an azimuthal electric field. An electron completing the upper traverse is accelerated, while the one on the lower traverse is retarded.

in section 1.5.2). A fundamental limitation of such an accelerator is the obvious one that acceleration, whether in such devices or the natural world, is necessarily pulsed or sporadic, since the magnetic field can neither increase nor decrease indefinitely. The betatron mechanism was suggested by Swann [12] as a possible particle accelerator at the Sun.

A phenomenon proposed by Sweet [13] as a mechanism to account for solar flares is that of magnetic annihilation. In this process, usually described in terms of magnetic fields rather than the currents which generate them, adjacent regions of opposing field coalesce to cancel each other out, wholly or partially, and the energy that had previously been contained in the fields is released to appear as kinetic energy of accelerated particles. Another way of looking at the same phenomenon is to see the charged-particle flow responsible for the fields becoming disrupted, and the energy stored in the field caused by the flow being re-distributed between these and/or neighbouring charged particles. Note that, however it is viewed, magnetic annihilation involves temporal change.

A closely related phenomenon is that of magnetic reconnection [14] (or often simply reconnection). When this is seen as an essentially temporal change, reconnection becomes synonymous with magnetic annihilation. There are, though, instances where reconnection is considered to take place as a steady-state merging of opposing fields with no temporal element. We shall discuss this, and the apparent, or real contradiction it represents, when we come to consider acceleration in the Earth's magnetosphere in Chapter 8.

1.2.4.4 Feedback

Temporal changes in magnetic field strength originate from changes in electric currents. Since currents, being essentially electric charge in motion, have their origin in the flow of charged particles, and such particles are, in turn, influenced by the magnetic field, the flow may well be modified in the process. A complex feedback between changing fields and particle acceleration is to be expected, just as in the electric counterpart.

1.2.5 Acceleration in combined electric and magnetic fields

The acceleration caused by electric and magnetic fields combined is given by the Lorentz force divided by mass, i.e.

$$a = e(\mathcal{E} + vB\sin\alpha)/m$$

where $\mathcal{E}$ is the combined space-charge and inductive electric field, and α is the pitch angle. The first component of the force is in the direction of the combined electric field, and the second is at right angles to both v and B.

It follows from previous discussion that a static magnetic field introduced into a space-charge region will influence the paths taken by charged particles, but will have no effect on their kinetic energies. If, on the other hand, the magnetic field strength or direction changes with time, this is no longer the case because the particle experiences, in addition to the space-charge field, an electric field induced by the changing magnetic field. Under these circumstances the combined electric field can no longer be represented conveniently by a set of equipotentials. In the case of a magnetic induction changing at a rate of one tesla per second, the kinetic energy gained (or lost, depending on the sense of motion) by a singly charged particle gyrating around a circle one square metre in area is one electron volt. If the rate of change of the field, the size of the area or the charge carried by the particle are increased, the kinetic energy is changed in the same proportion. We can now consider some examples.

A practical example of an accelerator dependent upon a static magnetic field and a time-varying electric field is the cyclotron [15]. In a cyclotron (see figure 1.12) charged particles gyrating in a magnetic field are accelerated twice per orbit, as they cross from one D-shaped electrode to the other. The particles are accelerated by an electric field which oscillates at the gyro-frequency. Acceleration is accomplished in this device in a large number of small steps. The process may be continued to relativistic energies if the oscillator frequency is reduced to allow for the fact that the gyro-frequency reduces as energy is gained. This is done in a synchrocyclotron [10]. Although these are clearly highly contrived devices, it is clear that even in the randomness of nature any electric field oscillating at the gyro-frequency of a particular species of particle has potential for delivering or removing energy. Electron and ion cyclotron waves, whose electric field directions rotate at the electron and ion gyro-periods, are very effective accelerators, or retarders, of particles which happen to be in phase with the wave, or in opposite phase. Such waves are employed to energise particles in the laboratory and are thought to have an equally important role in nature. Gyro-resonant acceleration is met in the everyday world in the act of rotating a skipping rope, or in the rather more spectacular acceleration of the hammer at the Olympic games.

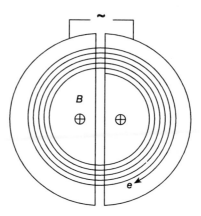

Figure 1.12. Principle of the cyclotron. Electrons, or other charged particles, are guided by a magnetic field to cross repeatedly a gap within which there is an electric field oscillating at the right frequency and phase to accelerate the particles between the D-shaped electrodes.

The combined effects of electric and magnetic fields in electromagnetic radiation, a spectrum extending from radio waves, via infra-red radiation, through visible and ultra-violet radiation, to x-rays and gamma rays, deserve specific mention. In electromagnetic waves the electric oscillations are inextricably linked to the magnetic oscillations. In free space electromagnetic waves travel at c and the electric field in V m^{-1} is 3×10^8 times the magnetic induction in T.

Electrons and other particles can absorb energy from electromagnetic radiation in a variety of ways. Electrons can be ejected from atoms by absorbing visible light or x-rays in the process called photo-emission. X-rays are scattered and partially absorbed in Compton scattering, and energy may be injected into free electrons by means of their sympathetic oscillation with radio waves, the waves themselves undergoing Thomson scattering.

The phenomenon of photoelectric emission is employed in the first stage of a photomultiplier or image intensifier [15, 27] where electrons released by visible, infra-red or ultraviolet radiation are accelerated by an electric field onto another surface (or dynode), or another part of the same surface. There they each knock-on several secondary electrons which are accelerated onto yet another (or another part of the same) surface. A cascade develops through several stages culminating in the deposition of a substantial negative charge at the detector's anode. The consequent drop in electric potential can be detected and counted, or the restoring current may be measured. Photoelectric emission by solar ultra-violet radiation causes spacecraft to become positively charged relative to the surrounding medium. In another example, laser beams are launched into media where their energy is deposited in the electrons as a stage towards heating the medium.

One of the fundamental properties of a charged particle is that, when it is accelerated in the most general sense, i.e. when its velocity is changed in magnitude or direction, the particle emits electromagnetic radiation. This is the principle upon which a radio antenna works. The power emitted by an electron undergoing acceleration a m s^{-2} is

$$P_{\text{radiated}} = 5.7 \times 10^{-54} a^2 \text{ W}.$$

The fact that power is emitted during acceleration sets a limit on the rate at which the energy of a charged particle can be increased. If it tries to take on energy too quickly, the energy is just radiated away. If we express acceleration a in terms of the rate of change of kinetic energy and equate the resulting expression to the rate of loss of energy by radiation, we can estimate the maximum rate at which an electron can take energy on board. The rate

$$(\mathrm{d}E/\mathrm{d}t)_{\text{max}} = 8.13 \times 10^{28} (1 + E/E_0)^4 ((E/E_0)^2 + 2E/E_0) \text{ eV s}^{-1}$$

is plotted in figure 1.13. It looks very large, and indeed will not be much of a limitation in the acceleration of electrons, in the terrestrial arena. Sometimes, though, even these large rates do represent a significant limitation. To illustrate this let us see how they apply to the acceleration of an electron of initial energy 1 eV in the electric field of an x-ray of 100 keV. The period for an oscillation of the electric field from a maximum in one direction to a maximum in the opposite direction of an x-ray of this energy is 1.5×10^{-19} s. If the electron were to absorb the full 100 keV in this time, the rate of absorption would be 6×10^{23} eV s^{-1}. The maximum allowable rate, according to figure 1.13 is only 3.2×10^{23} eV s^{-1}, so the effect would be for the electron to radiate energy as fast as it attempts to store it. The net result is that an observer would see simply an x-ray being scattered by an electron.

1.2.6 Summary

Electric and magnetic fields that change with time, or whose position changes with time are able to accelerate charged particles, as are electromagnetic fields which exhibit both of these attributes. Electric fields which have their origin in static space charge, and unchanging magnetic fields are not accelerators. Temporal change is an essential ingredient of an accelerating mechanism.

1.3 Assemblies of particles

So far we have been discussing the acceleration of individual particles. Such a treatment is adequate in applications where the particles to be accelerated all

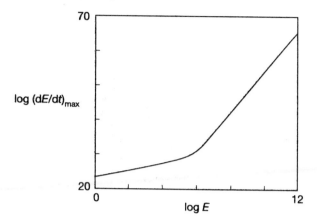

Figure 1.13. Estimate of the maximum rate at which an electron can be accelerated before immediately shedding by radiation all energy gained. The ordinate shows the log of the maximum rate of increase of kinetic energy in eV s^{-1}, and the abscissa the log of the energy in eV.

stem from a source whose spread in initial energies is negligible in relation to the energy gained during acceleration. The situation in space is different, however. The particle populations there generally have so wide a range of energies that any particular accelerator may affect different particles in different ways. We have no choice, therefore, but to consider, not just how individual particles behave, but also the wider question of how particles are distributed in space and velocity and how this distribution behaves.

1.3.1 Velocity distributions

To be able to describe completely all there is to know about a particle population at a specified place and time we need to know how many particles of any given species there are per unit volume and how they are distributed in velocity or momentum. Figure 1.14 shows what is involved. Figure 1.14(*a*) depicts a three-dimensional volume *V* of configuration or 'real space' within which there are several particles moving at a variety of speeds and directions as indicated by the arrows. The volume hosts three particles, **A**, **B** and **C**. **A** and **B** have equal velocities almost directly parallel to the *z* direction, while the third is moving parallel to the *x* axis at a slower speed, as indicated by its shorter arrow. If the sides of *V* are each of length 1 m, we can start to specify the distribution by noting that the number density is 3 m^{-3}.

Figure 1.14(*b*) shows another three-dimensional space, organised this time by a grid of velocities. The axes are velocity along the *x* direction v_x, and velocities in the *y* and *z* directions v_y and v_z. It shows the three particles **A**, **B** and **C** in the positions appropriate to their velocities. **A** and **B** lie within the

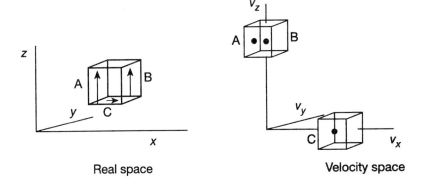

Figure 1.14. Velocity-space density and phase-space density. To the left, three particles in the same element of real, or configuration space have the velocities shown. **A** and **B** have a common velocity which is directed parallel to the z axis, while **C** moves more slowly in the x direction. To the right, these three particles occupy two elements of a velocity space whose axes represent velocities in the x, y and z directions. **A** and **B** appear in an element on the z axis, while **C** is on the x axis. Velocity space density is defined as the number of particles per unit 'volume' of velocity space per unit volume of real space. If momentum is used instead of velocity for the axes of the left-hand part of the figure, the resulting number of particles per unit volume of momentum space per unit volume of real space is the phase-space density.

'volume', $\mathcal{V}_1$ clinging to the v_z axis, and **C** finds itself in $\mathcal{V}_2$ impaled on the v_x axis, closer to the origin. There are no arrows in (*b*), because the speeds and directions are already specified by positions in this part of the figure. If the sides of $\mathcal{V}_1$ and $\mathcal{V}_2$ are each of 'length' 1 m s^{-1}, the respective densities in velocity space are 2 m^{-3} s^3 and 1 m^{-3} s^3. Similarly, if figure 1.14(*b*) were constructed with axes of momentum, rather than velocity, and the sides of the corresponding 'volumes' $\mathcal{V}_1$ and $\mathcal{V}_2$, therein, were 1 N s, the respective densities in momentum space would be 2 (N s)$^{-3}$ and 1 (N s)$^{-3}$.

1.3.1.1 *Velocity-space density*

Number densities in real space and velocity space can be combined into a (confusingly named) velocity-space density.

Velocity-space density, f, is defined as the number of particles per unit volume of real space per unit volume of velocity space, i.e.

$$f = N/V\mathcal{V}$$
$$= N/\mathrm{d}x\,\mathrm{d}y\,\mathrm{d}z\,\mathrm{d}v_x\,\mathrm{d}v_y\,\mathrm{d}v_z.$$

We shall give the product $V\mathcal{V}$ the symbol $\mathcal{T}$.

In the example of figure 1.14 the velocity-space density at the velocities of **A** and **B** is 2 m^{-6} s^3. At the velocity of particle **C** it is 1 m^{-6} s^3. In order to avoid very small numbers later we shall measure velocity-space density throughout in units of km^{-6} s^3. A census of velocity-space densities at every point in velocity space provides a complete description of a particle distribution at a given location.

1.3.1.2   *Phase-space density*

Number densities in real space and momentum space can be combined into an equivalent (obscurely named) phase-space density.

Phase-space density, F, is defined as the number of particles per unit volume of real space per unit volume of momentum space:

$$F = N/V\mathcal{V}$$
$$= N/dx\,dy\,dz\,dp_x\,dp_y\,dp_z.$$

The product $V\mathcal{V}$ is the volume in phase space. We adopt for it the symbol τ.

In the example of figure 1.14 the phase-space density for particles of the momenta of A and B is 2 (J s)$^{-3}$. At the momentum of particle C it is 1 (J s)$^{-3}$. In order to avoid very large numbers later we shall measure phase-space density throughout in units of (zJ s)$^{-3}$. A census of phase-space densities at every point in momentum space provides a complete description of a particle distribution at a given location.

For non-relativistic electrons

$$F \left[\text{in (zJ s)}^{-3}\right] = 1.32 \times 10^9 f \left[\text{in km}^{-6}\text{ s}^3\right].$$

For non-relativistic protons, F is smaller by a factor of 6.2×10^9 for the same f.

For relativistic electrons, whose velocities increase only imperceptibly with energy, velocity space and velocity-space density cease to be useful concepts. Momentum, however, being unbounded, increases without limit as energy increases, as shown in figure 1.15, making phase-space density a useful measure at all energies.

Velocity-space density and phase-space density will prove to be invaluable measures of particle presence in what follows. When all else is literally in a state of flux, these quantities will serve as guiding beacons.

1.4   Generation of velocity distributions

So far, we have discussed velocity and momentum distributions purely in the abstract, without indicating anything about their actual shape or form. Would we expect, for example, to find velocities clustered in a narrow range, or are they likely to be widely spread? Are low velocities more common than high velocities, or vice versa? Let us look into this now.

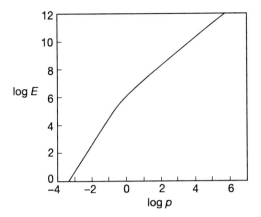

Figure 1.15. The relation between energy (in eV) and momentum (in zN s) for electrons. At low velocity, energy is proportional to the square of the momentum. At high, relativistic, velocities, energy becomes directly proportional to momentum.

Since the primordial distributions from which others later evolve are generally the result of random processes, we shall begin with a discussion of what is likely to arise purely by chance.

1.4.1 Entropy

A question that poses itself is: 'what is the most likely distribution to be taken up by a population of a fixed number of particles able to exchange at random, by means of collisions, for example, all or part of their energy with others?' This might at first seem an impossible task. It would be, without our consistent experience that in any random process, which of course these uncoordinated collisions amount to, there is always a tendency for extremes to disappear and for things to level out. Consider for example some small heaps of sand on a tray. Shake the tray, and note that the heaps disappear and a level distribution is reached. Using children's bricks construct an intricate arrangement of bridges, towers, etc, and invite a young child to 'inspect' them.

The above tendency, expressed formally, is the tendency for a property of any system, known as entropy, S, to increase as events proceed. A measure of entropy, which derives from Boltzmann, is (see, for example, [16])

$$S = -\sum_i k N_i \ln F_i + \text{constant}$$

where k ($= 1.38 \times 10^{-23}$ J K^{-1}) is Boltzmann's constant, N_i is the number of particles in a state, or cell i, and F_i is the phase-space density for cell i. The sum is taken over all cells. Since it is only changes in entropy that interest us, the above constant may arbitrarily be set to zero, so the quantity could equally well be expressed in terms of velocity-space density. The appendage of an arbitrary constant is a property entropy shares with potential. Since it is, furthermore, only the sense or direction of change that will be important to us here, we can even ignore the magnitude of the multiplying constant k, and can choose freely whether to work in Naperian logarithms, or natural logarithms as here. We shall now find for ourselves the power and usefulness of this expression. Note the minus sign.

1.4.2　Level or flat distributions

Imagine that there are just two possible states, or 'cells' of real/velocity space T_1 and T_2 for $N_1 + N_2$ particles to occupy. These six-dimensional volumes are represented in figure 1.16 in one dimension. To start with, put N_1 in T_1 and N_2 in T_2. Now let the particles be charged and be randomly jostled by an agency such as a time-dependent electric field, in such a way that a given particle has at any time an equal chance of finding itself anywhere within the combined volume $T_1 + T_2$. Note that if occupants of the cells have different energies, which in general they will, the agent will need to provide or remove energy from the particle population. How many particles would we expect to find in T_1 and T_2 if we took an instant census after a really good shake-up?

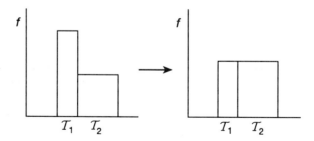

Figure 1.16. Levelling, and entropy increase, produced by random jostling of an initially non-uniform distribution of velocity-space density.

For the answer we need to find the state giving the maximum value of entropy. That is we need to find the maximum value of

$$S = -N_1 k \ln(N_1/T_1) - N_2 k \ln(N_2/T_2)$$

while keeping the total number of particles, $N_1 + N_2$, constant.

Found either by differentiation or by trial and error the unsurprising result is that the numbers found in the two cells are simply proportional to the volumes of the cells, so that

$$f_1 = f_2.$$

So, when the *a priori* probabilities are equal for equal volumes, one-dimensional velocity-space density becomes uniform. Further jostling produces no further change. The same applies to any number of dimensions. It applies too for any number of cells. Changes of entropy as particles are moved from one cell to another will increase entropy only if they produce a flatter distribution. As a result we can anticipate a distribution that is uniform over all cells, just as in the three-dimensional analogy of the number density of gas molecules in a room.

1.4.3   The Landau plateau

Circumstances can and will arise where the external agency is able to jostle only those particles within a limited range of velocities, i.e. where there has to be resonance before kinetic energy can be changed. An often spectacular example of a resonant interaction is that between an ocean wave and a surfer. Surfing is possible only when the surfer catches a wave by ensuring that when a wave arrives, he or she is already travelling close to the wave speed. The larger the wave the greater is the tolerance on this initial resonance. Would-be surfers travelling too fast are not caught by the wave and those moving too slowly ride over it and are left behind. The same principles apply to charged particles in a medium permeated by waves of electric potential, electrostatic waves. If the waves are all of the same velocity or are of a limited range of velocities, random changes or 'random walk' in velocity are undertaken only by charged particles with resonant velocities. Consequently, the levelling effect of the increase in entropy is confined to a limited range thus creating a flat region or 'plateau' in the velocity distribution of the particles as indicated in figure 1.17. Higher and lower velocities are left unchanged. This effect, which will feature strongly in much of what follows, was foreseen by Landau [9]. The attenuation suffered by the waves in creating the plateau is known as Landau damping.

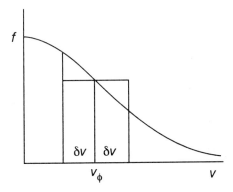

Figure 1.17. Formation of a plateau (horizontal line) in a velocity-space distribution of charged particles subject to random jostling by electrostatic waves, as foreseen by Landau. The position of the plateau is determined by the phase velocity, v_ϕ and its extent, $2\delta v$ by the amplitude of the waves.

Our studies carried out in connection with the aurora in Chapter 3 suggest that the plateau need not develop systematically or monotonically, but that the entropy increase might allow at first the formation of a sharp peak. The computer code in Appendix 3 invites you to check this for yourself.

Are distributions always destined ultimately to flatten out over the full or partial range, or are there other forms of equilibrium distribution?

1.4.4   The Maxwell–Boltzmann distribution

We now consider what will happen if particles of a gas collide with each other at random and in so doing exchange some or all of their energies in the process. This is similar to the example considered above except that now the jostling is due entirely to collisions between particles and there is now no external supply or sink of energy. Particles can gain energy only at the expense of others. What is the distribution giving maximum entropy now? Specifically, how would a fixed number of particles divide themselves among a set of cells corresponding to different energies? There are many ways of looking into this. The analytical treatment of Maxwell and Boltzmann is a standard element of statistical mechanics [16]. A purely numerical method, made possible by the advent of computers, begins by allocating a chosen number of particles at will among a series of adjoining cells which for simplicity are equally spaced in energy. The entropy of this state can be evaluated from

$$S = -N_1 k \ln(N_1/T_1) - N_2 k \ln(N_2/T_2) - N_3 k \ln(N_3/T_3) - \ldots .$$

To explore, move one particle from one of these cells to the next higher energy cell and move a particle from this (or any other) cell down in energy to compensate. Re-assess the entropy. If the entropy has decreased, disallow the change and move the particles back to their places. If entropy has increased, leave the particles in their new cells, select a new particle at random, and perform the test again. Keep going until you find a distribution among the cells which cannot be raised in entropy by any change of this kind. The resulting distribution is the one we are after. It would obviously take quite a while to complete this exercise by hand, but it is worth trying a few iterations to discover the trend. A computer, however, makes investigation by these means wholly practicable. A computer code for this exercise appears in Appendix 1, and some results are shown in figure 1.18. This figure shows the evolution of f, as a distribution concentrated initially in cell 2, representing two units of energy, rearranges itself according to the 'rules' of conservation of number and energy and the increase of entropy. The result is that the distribution evolves into

$$f(E) \propto \exp(-E/E_M)$$

where E_M is an energy characterizing the distribution. In this one-dimensional analysis, it is also the mean energy. Once the distribution has attained this form, which appears as a straight line in a log-linear plot, there is no further change. A similar distribution, with appropriate E_M, results whatever the initial distribution. A second example with an initial and mean energy of three units is shown in figure 1.19. If a distribution of this form is chosen initially, there is no evolution.

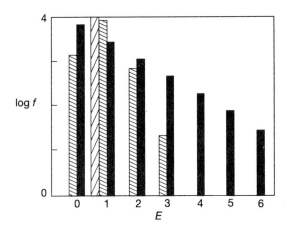

Figure 1.18. Result of a numerical experiment in which constituents of an initially mono-energetic distribution (lightly hatched column at $E = 1$) are subject to random jumps in energy subject to the constraints that the total number of particles and the total energy remain constant. The energy constraint is imposed in the experiment by compensating energy increases by one particle with an equal loss by another. The distribution approaches, via a series of intermediate distributions (e.g. the heavily hatched one), an exponential form (black) after which there is no further change. The final distribution is the Maxwell–Boltzmann distribution.

In the above we have decreed that entropy should increase. What happens if we just allow nature to take its course? The computer code in Appendix 2 does just this. It selects two particles at random from an initial distribution and allows them to exchange all or part of their energy, the only stipulation being that their energy jumps are of the same magnitude and in opposite directions, thus conserving energy. We find that entropy automatically increases and the distribution still evolves into a Maxwellian, even without a helping hand. Experimentation with this code will reveal many facets of this fundamental process, such as the rate of progress and the statistical fluctuations that continue to occur about the 'equilibrium' state.

The full six-dimensional analysis shows that if the total number density (which remains constant throughout) is n m^{-3}, and E and E_M are measured in eV,

$$f(E) = 90n E_M^{-3/2} \exp(-E/E_M) \text{ km}^{-6} \text{ s}^3.$$

The mean energy $\langle E \rangle$ $(= 1.5 E_M)$ also remains constant, since energy is neither

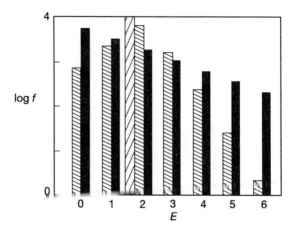

Figure 1.19. The same as figure 1.18 but with the initial mono-energetic distribution at $E = 2$. Note the flatter equilibrium distribution.

added nor taken away. The distribution is known after its discoverers as the Maxwell–Boltzmann distribution and is one of the most important in all physics.

E_M (in J), divided by Boltzmann's constant k is known as the temperature, T (in K). With E_M and $\langle E \rangle$ (in eV) we have

$$T = 1.16 \times 10^4 E_M \ \text{K}$$
$$= 7.74 \times 10^3 \langle E \rangle \ \text{K}.$$

Another way to discover the equilibrium distribution (again for simplicity in one dimension) is to follow an algebraic method. Consider two pairs of cells, the members of each pair being separated by the same energy step δE. Let the numbers, volumes and velocity-space densities be $N_1 \ldots N_4$, $T_1 \ldots T_4$ and $f_1 \ldots f_4$. Perturb the distribution by moving N particles from 1 to 2 and the same number from 4 to 3, thus preserving particles and energy as required. Since we are considering only a minor perturbation, we arrange for N to be very much smaller than any of the numbers $N_1 \ldots N_4$. We find that change in entropy is

$$\frac{\Delta S}{k} = N(\ln f_1 - \ln f_2) + N^2 \left(\frac{1}{N_1} + \frac{1}{N_2} \right) + N(\ln f_4 - \ln f_3) - N^2 \left(\frac{1}{N_3} + \frac{1}{N_4} \right)$$

or

$$\frac{\Delta S}{k} = N \ln \left(\frac{f_1 f_4}{f_2 f_3} \right) - N^2 \left(\frac{1}{N_1} + \frac{1}{N_2} + \frac{1}{N_3} + \frac{1}{N_4} \right)$$

whatever the values of $T_1 \ldots T_4$.

If the distribution $f_1 \ldots f_4$ is in equilibrium, i.e. stable against this type of perturbation ΔS must be negative for both possible signs of N. The second term is always negative, but the first changes sign with N. The only way we can be assured of a negative ΔS, therefore, is to make the first term zero, i.e.

$$\frac{f_1}{f_2} = \frac{f_3}{f_4}.$$

The distribution would be in equilibrium, therefore, if the same step in energy were met by the same factor change in velocity-space density. In other words,

$$f \propto \exp(E/E_M)$$

as found above.

Figure 1.20 shows two Maxwellians, both drawn for an electron number density of 1 m^{-3}. One has a temperature of 2×10^4 K and the other is twice as hot. Figure 1.20 (*a*) uses linear scales for both f and E; (*b*), with a log scale for f and a linear scale for E reveals the distributions in their simplest form; while (*c*) has logarithmic scales for both axes to demonstrate that in this often-used framework all Maxwellians are geometrically similar, regardless of temperature or density.

The concept of temperature applies only to an assembly whose energy distribution is a Maxwellian. Sometimes the concept is extended to non-Maxwellians of known or measured mean energy by quoting an effective temperature T_{eff} defined, for $\langle E \rangle$ in eV, as

$$T_{eff} = 7.74 \times 10^3 \langle E \rangle \text{ K}.$$

While this may well be useful in some circumstances, it has to be treated with some caution. The assumptions must continue to be kept in mind, since mean energies evaluated from only part of a distribution, as is invariably the case, may not be fully representative. After all, a mean energy so determined is bound to lie within the energy range of measurement, whereas the true mean could conceivably lie outside.

Since it is the constraint of the total energy being conserved that prevents the distribution becoming progressively flatter, the Maxwellian may be considered to be the result of the levelling of a distribution as far as it can go with a given particle and energy content.

1.4.5 The power law

Although Maxwellians are ubiquitous in everyday gases and particle assemblies where collisions are the primary form of energy exchange, they are not common in space where other factors dominate. Distributions found in space are, as we shall see, often well represented over much of their energy range by a power law or near power law. It is of interest therefore to consider what general influences might lead to these and other types of distribution. It should be said at once

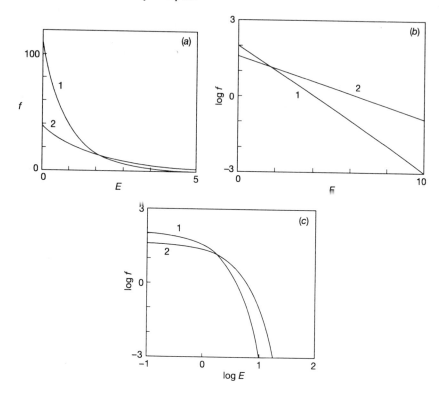

Figure 1.20. Maxwellians of the same density but different temperatures, the temperature of **2** being twice that of **1**. (*a*) uses linear scales for both velocity-space density f and energy. The straight lines in (*b*), where f is on a logarithmic scale, reveal the exponential nature of the distributions and, through their slopes, the relative temperatures. It can be seen from (*c*), where both scales are logarithmic, that, in these terms, the shape of a Maxwellian is invariant—the location along the energy axis being determined by the temperature, and location along the velocity-space-density axis by the number density.

that this is not an easy task, and indeed there is little in the form of established theory to go on. However, there are some promising leads to explore.

A mechanism for producing a power law was invented by Enrico Fermi. In his celebrated cosmic ray theory [2] Fermi combined three powerful ideas. Firstly, he envisaged the galaxy as containing 'wandering magnetic fields' which would serve as scattering and accelerating centres. He noted, secondly, that embryonic cosmic rays would randomly gain and lose energy from head-on and overtaking collisions which, at the relativistic speeds involved, would be in amounts in proportion to their current energy. Secondly, he pointed out that particles would on average experience more head-on (energy-gaining) collisions than overtaking (energy-losing) collisions for the same reason that a car windscreen becomes wetter than the rear window. As a result, cosmic rays would make a net gain of energy with time. The third factor was to envisage a loose, statistical time limit to the process, arising from ever-present possibilities of escape from the system or of an energy-sapping collision with another galactic particle. He thus envisaged a 'lifetime' which allowed fewer and fewer particles to reach greater ages and consequently higher energies. As described in Chapter 10, the result of such a process is that

$$f_{UR}(E) \propto E^{-\gamma}$$

where the exponent γ depends on the ratio of the mean time between each gain or loss, on the characteristic escape time, and on the speed of the scattering centres.

Following Fermi's basic ideas, Davis pointed out [17] that even without systematic energy gain the inevitable increase in entropy during a random walk in energy would broaden the distribution and thus accelerate some of the particles. The losses or escape considered by Fermi would ensure that as a given injection of particles broadened in energy the numbers would diminish. The overall result from a continuous injection would be a combination of relatively large numbers of fresh particles with narrow distributions and relatively few old particles with a broad distribution, together with all stages in between (echoing the human condition?). Davis treated the problem as one of diffusion, which is tantamount to assuming increasing entropy.

The combination of entropy increase and exponential decay also leads to a power law, as we found, too, using a statistical approach [18], the exponent depending on the same key factors as in Fermi's analysis of the systematic energy gain. Using the general expression for energy gain given earlier, the analysis has here been extended down to include non-relativistic energies. The general formula for random energy gain in acceleration by this 'dynamic scattering' is given in Appendix 4 and two examples showing the change in spectral index at the transition from non-relativistic to ultra-relativistic energies (around log $E = 6$) are shown in figure 1.21.

We shall have recourse to these various distributions when discussing solar-flare electrons in Chapter 9 and cosmic rays in Chapter 10.

1.4.6 The kappa distribution

A major snag with any true power-law distribution is that it cannot really exist in practice. If γ is 1 or more, the total number density becomes infinite due to the increasingly large numbers of particles at lower and lower energies. If γ is 1 or less, the total energy density becomes infinite because of excessive numbers at high energies. It is clear, then, that even when a set of measurements yields a power law, the true distribution must certainly depart from this outside the measured range.

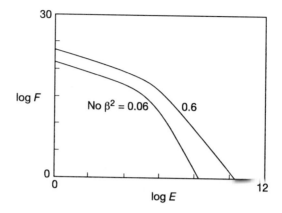

Figure 1.21. 'Dynamic-scatter' distributions resulting from random energy exchanges between particles and moving scattering centres. Energies range from sub-relativistic to ultra relativistic. The form of the distribution depends on the product of the average number of energy exchanges N_0, with the square of the speed of the scattering centres relative to the speed of light β^2. Curves are shown for two values of this product. The power-law form of the distributions at relativistic energies was deduced by Davis in connection with Fermi's theory of cosmic-ray acceleration. The transition to a lower slope at non-relativistic speeds is a new result.

The kappa distribution [19] offers an escape. This distribution is actually a whole family of distributions. The family members complete a continuous transition from Maxwellians at one extreme to power laws at the other. Moreover, each member of the family looks like a Maxwellian at low energies, and a power law at high energies. The kappa distribution is best represented for immediate purposes as

$$f(E) \propto (E + K)^{-\gamma}.$$

It could be described as a modified power law, i.e. a power law in energy plus a fixed energy K. The distribution takes its name from an index κ which gives rise to the exponent through

$$\gamma = \kappa - 1.$$

There is a characteristic energy E_κ where

$$E_\kappa = K/\kappa.$$

The different members of the family are found by entering into the above formulae values of κ and E_κ ranging independently from zero to infinity. If either of these is zero then K is zero, and the distribution becomes a power law of negative exponent $\kappa - 1$. If $\kappa \gg 1$, the distribution becomes a Maxwellian of characteristic energy E_κ. Normalization details are given in Appendix 4.

Figure 1.22(*a*) illustrates this transformation in a framework in which a power-law is a straight line. Figure 1.22(*b*) does the same in a framework in which the kappa distribution is a straight line.

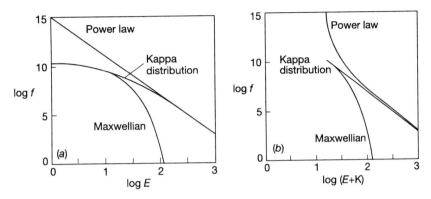

Figure 1.22. Comparison of Maxwellian, power-law and kappa distribution. The kappa distribution takes the form of a Maxwellian at low energies and becomes a power law at high energies. (*a*) shows log *f* plotted against *E*; (*b*) log(*E* + K) is given as abscissa to illustrate that the kappa distribution takes the form of a negative power law in *E* + K.

The kappa distribution will be found to be helpful later when discussing electrons of the solar wind.

1.4.7 Composites

When a measured distribution cannot be represented at all satisfactorily by a standard such as a Maxwellian, power-law or kappa distribution, recourse is sometimes made to a composite—usually the sum of two separate Maxwellians or two power laws. This can sometimes be a convenient way of summarizing a distribution. If the distribution is anisotropic, this approach may also be helpful in conveying the fact that there are different 'temperatures' in directions parallel to and perpendicular to a key direction such as that of the magnetic field. In this latter case, the composite is referred to as a bi-Maxwellian. If, on the other hand, a composite distribution is seen as more than a convenient mathematical representation, and is taken to give insight into the underlying physical processes, it could be seriously misleading. This is because particles from different sources cannot, when acted upon by the same set of prevailing electric and magnetic fields, be brought together into the same region of phase space at the same time, any more than two identical balloons can be brought into collision by air currents. There is an illogicality, too, in considering particles inhabiting the same region of phase space, i.e. particles that are indistinguishable in all respects, as belonging to and behaving differently as members of different families. The logical inconsistency of such an interpretation is exposed by figure 1.23.

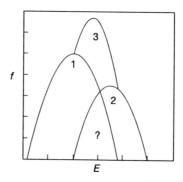

Figure 1.23. Dilemma posed by considering a velocity-space distribution to be the merging of two component distributions. If distribution **3** is the sum of distributions **1** and **2**, the particles in the region of overlap are identical in every respect yet are required to have followed different trajectories under the influence of the same forces!

Due, however, to the fact that all observations are made over finite intervals of time and over finite volumes of phase space, it is possible for more than one family of particles to share a 'volume' and time interval and to appear thereby to combine into a single distribution. Possible distributions arising from **1** and **2** in figure 1.23 would, without modification to either, lie along or below, but not above, their envelope. The cosmic-ray power laws discussed above are of this type.

A statement that will help us treat this and other questions of continuity in a formal way—in fact one of the most far reaching in the whole of physics, Liouville's theorem—is about to join our armoury.

1.5   Metamorphosis

How can velocity distributions, once formed, become altered?   How, for example, can a Maxwellian, once established, ever be changed? Is it possible for a monotonic distribution acted upon by suitable forces to develop one or more peaks in velocity-space density? Can any nominated change be effected? What are the constraints if we restrict ourselves to purely dynamical or reversible changes? How much is the position relaxed if there are irreversible happenings, such as energy dissipation, ionisation or radioactive decay?

1.5.1   Liouville's theorem

A universally important and far reaching statement on what continuity permits and what it disallows is provided by Liouville's theorem [16] and in particular, its application to electromagnetic forces [20]. This theorem states that velocity-space density and phase-space density remain unchanged along any path followed in response to dynamical forces, i.e. along any path that would be retraced if particles following it were stopped and sent back with velocities reversed. We can now see, by means of some specific examples, what this theorem tells us and how it constrains the changes that can occur.

Liouville's theorem can be demonstrated pictorially in two dimensions. In a 'space' of ordinate velocity v and abscissa position coordinate x, we draw a rectangle of sides δx and δv, as shown in figure 1.24, and place within it N particles. We could start with any shape whatsoever, but the rectangle is the most convenient. The two-dimensional 'volume', $\mathcal{T}$, is $\delta x \delta v$, and the two-dimensional velocity-space density f is $N/\delta x \delta v$. We now let the particles move at their endowed velocities for a while. It does not matter how long as long as it is the same for all. After a time the particles will have moved to different positions. They will all have moved parallel to the x axis since their velocities have not changed. Particles travelling at the same speed will remain the same distance apart. The faster ones will have moved further than the slower ones, so the particles will have become more spread out, deforming the original rectangle into a parallelogram. Since the base and height of the parallelogram are the same as the base and height of the original rectangle, the area, representing the two-dimensional volume, will also be the same. The velocity-space density will, as a consequence, also be unchanged, since there are still the same number of particles in the same 'volume'. It is clear, too, that if we replace the original rectangle by any shape whatever, or even any combination of shapes, we will always find that the velocity-space density remains constant, since we can consider each original shape to be composed of a number of small rectangles which will all behave in the same way. By similar reasoning, this thought experiment can be performed in four or the full six dimensions.

What happens if the particles do not travel a constant speed but are accelerated *en route*? Let us begin with an acceleration a applied to all particles for the same time, t. The infinitesimal distance $\mathrm{d}x$ travelled in each infinitesimal time $\mathrm{d}t$ is

$$\mathrm{d}x = v\,\mathrm{d}t$$

in the process of which the velocity increases infinitesimally by

$$\mathrm{d}v = a\,\mathrm{d}t.$$

As time progresses, test particles at the corners of the rectangle in figure 1.25 will move along the paths indicated. which have been computed step by small step to illustrate this development

The overall effect is that all velocities have increased, in this case, by four units on our arbitrary scale. In order to find the new 'volume' we again need to know the base and height of the new parallelogram. Since particles of the same initial speed will throughout have the same speed as each other, they will remain the same distance apart from beginning to end, making the base of the parallelogram the same as the base of the original rectangle. The same applies

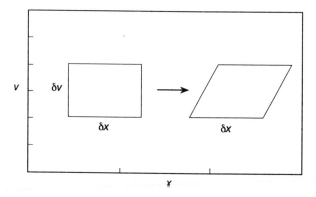

Figure 1.24. Illustration in two dimensions of the conservation of velocity-space density predicted by Liouville's theorem. A set of particles occupying initially the rectangular 2d 'volume' on the left will, after the passage of time, occupy a parallelogram whose base has moved forward a shorter distance than its top. Since base and height are both preserved, the 'volume' and velocity-space density are unchanged.

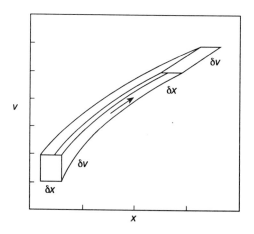

Figure 1.25. Illustration in two dimensions of the conservation of velocity-space density for particles undergoing uniform acceleration. The parallelogram resulting from the passage of time will, despite the increase in all velocities, maintain the same base length and height as the original rectangle, thus ensuring that velocity-space density is conserved under these conditions, too.

to the top, of course. Since acceleration for a given time increases all speeds by the same amount, velocities of particles at the top of the rectangle will always exceed those of particles at the bottom by the same amount, thus keeping the height of the figure the same. The product of base and height, i.e. the 'volume', will therefore be unchanged, thus conserving velocity-space density.

1.5.2 Some consequences

1.5.2.1 For orbital mechanics

Consider two spacecraft orbiting the Earth, one following the other along an eccentric orbit. As they travel inwards from apogee towards perigee they will be increasingly accelerated, the leading one experiencing the increases before the other, causing the space between them to widen the closer they get to the Earth. The reverse will, of course, occur on the outward leg of the orbit, bringing the spacecraft back to their original spacing when apogee is again reached. If the spacecraft are not separate objects but are joined together, as is the case, for example, for a comet passing close to the Sun or a planet, the tendency for one edge to pull away from the other sets up forces (tidal forces), which under some circumstances cause the object to break up. Note, too, the necessity for a train of cars on a roller coaster to be joined firmly together, to combat the tendency described by Liouville's theorem for them to repeatedly separate when accelerating.

1.5.2.2 Collimation

One of the consequences of the constancy of velocity-space density is the narrowing of directions, or 'collimation' of a velocity distribution under acceleration. In figure 1.26 we see the now familiar box in velocity space, understood to contain a number of particles. If we accelerate the particles in the x direction, v_x will increase and they will occupy a new box shifted along the v_x axis as shown. The greater the acceleration, or the longer its duration, the greater will be the displacement. The new box has the same dimensions in the v_y and v_z directions as before, since v_y and v_z are unchanged. Since all velocities in the x direction have increased by the same amount, as in figure 1.26, the 'volume' is unchanged. The number of particles has not changed either, so velocity-space density is conserved. What has happened, though, is that the particles are now moving within a more restricted range of directions, as shown by the radial lines. The stronger the acceleration, the greater is this collimation.

1.5.2.3 Adiabatic compression

In order to appreciate further the significance of the conservation of velocity- and phase-space density under the influence of purely dynamical forces let us try to defy it.

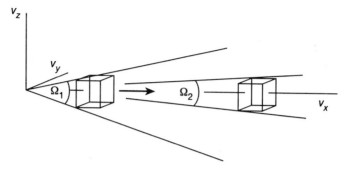

Figure 1.26. Collimation produced by acceleration. Particles travelling initially in a cone of directions within solid angle Ω_1 about the x axis congregate into the smaller solid angle Ω_2 when accelerated in the x direction.

We shall attempt to increase velocity-space density by reducing the size of a vessel containing a fixed number of particles. This will clearly increase the density in real space so might at first seem well worth a try.

We begin in a world of one spatial dimension and one velocity dimension. The volume will be reduced by means of a piston, as shown in figure 1.27. The speed of the piston is such that the length of the vessel reduces from L_1 to L_2 in an elapsed time t. The speed of the piston is, of course, simply the distance travelled divided by the time taken, i.e.

$$v_{\text{piston}} = (L_1 - L_2)/t.$$

We imagine the vessel to contain a number of particles, all of the same velocity, initially v_1 (which could equally well be thought of as a narrow range of velocities centred on v_1). Since this is a one-dimensional world, the particles can move only parallel to the sides of the vessel. We know from our earlier studies that each time a particle strikes the moving piston its velocity will increase by twice the speed of the piston, i.e. by

$$\delta v = 2(L_1 - L_2)/t.$$

The number of collisions is the total distance travelled (i.e. the average speed multiplied by the time), divided by the average length of the vessel. If the final velocity is v_2, the number of collisions is, then

$$N_{\text{coll}} = \frac{\frac{1}{2}(v_1 + v_2)/t}{\frac{1}{2}(L_1 + L_2)}.$$

To obtain the overall change in particle velocity we multiply δv by the total number of collisions, i.e.

$$v_2 - v_1 = N_{\text{coll}}\delta v$$
$$= (L_1 - L_2)(v_1 + v_2)/(L_1 + L_2)$$

or, writing L_1 / L_2 as the compression ratio C, and rearranging a little,

$$v_2 = Cv_1.$$

In one real-space dimension, then, the velocity of any particle and all particles will increase in direct proportion to the reduction in volume of the vessel. This would also be the case if we started with a distribution of particle velocities. At non-relativistic velocities, energy, being proportional to velocity squared, will increase by the square of the compression ratio, i.e.

$$E_2 = C^2 E_1.$$

It follows automatically that velocity ranges will also increase by the compression ratio, because both the lower limit of the range and the upper limit of the range will increase and therefore draw further apart. Another important consequence is that the process has added energy to the particle population, the energy having been drawn from the agency responsible for forcing the piston inwards against a battering from particles resisting the compression.

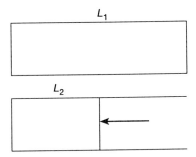

Figure 1.27. An attempt to defy Liouville's theorem by compression in one dimension.

These results are independent of the time taken, because if t is shortened there will be fewer collisions but each will produce a greater change of velocity. If the time is lengthened each collision will be less effective but there will be correspondingly more of them.

So what has happened to velocity-space density? Well, the number of particles will have remained constant, of course, while the volume of real space will have been reduced by the compression ratio, but the volume of velocity space will have increased by the same factor, through the widening of the range of velocities. The combined volume, T, will have remained constant. Figure 1.28, drawn for a compression ratio of 2:1, shows this pictorially. Velocity-space density and phase-space density are therefore unchanged, exactly as Liouville's theorem predicts.

An initial Maxwellian of just one dimension of velocity would be transformed from

$$f_1(E) \propto (n/E_M) \exp\{-E/E_M\}$$

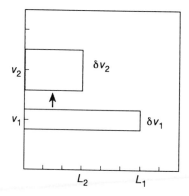

Figure 1.28. Compression in real space leads to a compensating expansion in velocity space. As the length of the cylinder reduces from L_1 to L_2, the range of speeds Δv increases by the same factor, thus preserving velocity-space density.

to

$$f_2(E) \propto (Cn/C^2 E_M) \exp\left\{-E/C^2 E_M\right\}$$
$$\propto (n/C E_M) \exp\left\{-E/C^2 E_M\right\}$$

i.e. a distribution with number density n raised by the compression ratio and all energies and consequently the temperature, raised by the square of the compression ratio as shown in figure 1.29. In this example with just one velocity dimension, the temperature, $C^2 E_M/k$, is inversely proportional to (real-space) volume squared, i.e.

$$T \propto V^{-2}.$$

It goes without saying that if we repeated the whole exercise, expanding the vessel instead of compressing it, the temperature would fall by the square of the expansion ratio.

What has happened, then, is that all velocity-space densities have remained the same but the particles composing them have moved, in the case of compression, to higher energies. The maximum velocity-space density n/E_M remains unchanged. Compression has led to a broadening of the distribution, to a rise of temperature, i.e. to 'heating' of the particles. The same happens to any form of original distribution. The distribution simply becomes broadened, which is represented on a plot with a $\log E$ abscissa by a shift of the whole curve sideways by a uniform factor towards higher energy.

Let us pursue this investigation into a four-dimensional space of two spatial and two velocity dimensions. Consider an assembly of charged particles in a narrow range of velocities gyrating in a magnetic field B. Instead of using pistons to reduce the volume, we shall increase the strength of the magnetic field, as in a betatron. The radius of gyration will reduce (see figure 1.30) in direct proportion

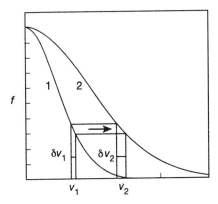

Figure 1.29. Change in velocity-space distribution under the compression illustrated in figure 1.27. The initial distribution expands as all speeds are increased by the same factor (by a factor of two in this example).

to the field increases. The path followed will be a very gradual spiral, like the track on a gramophone record or compact disc. The instantaneous radius of curvature will be, as we have seen, inversely proportional to field strength, and proportional to the particle's current momentum.

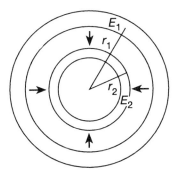

Figure 1.30. Compression in two dimensions of real space. Particles gyrating in a magnetic field initially in the outer ring of mean radius r_1 have energies centred on E_1. When the magnetic field is increased the mean radius reduces to r_2 and the energy is increased to E_2. It is shown in the text that the reduced volume in real space is compensated by a widening of the range of particle speeds, so that velocity-space density is conserved.

If such a transformation caused by gradually increasing field is followed, carefully, through gradually reducing radius and gradually increasing momentum it is found that a non-relativistic particle's kinetic energy will increase to a good approximation in direct proportion to the field strength. This is one of Alfvén's 'adiabatic approximations' which, as we shall see in Chapter 2, considerably simplify the computation of charged particle motion in a magnetic field. The term adiabatic in this context signifies that change must be gradual (relative to the gyro-period) for the approximation to hold.

The net effect of the links between energy, radius and field strength ensure that, if the initial radius of curvature is r_1 and the final radius has shrunk to r_2, under a linear compression ratio $C = r_1/r_2$ the final energy E_2 is related to the initial energy E_1 by

$$E_2 = C^2 E_1.$$

The range of energies will also increase by the factor C^2. The radius and the width of the ring in real space will both reduce by C, making the two-dimensional real-space volume, which is the product of these, also reduce by C^2. The combined real-space velocity-space volume $\mathcal{T}$, will therefore remain constant, as Liouville's theorem assures us it will.

This means that compression in two spatial dimensions leads to an increase in energy in direct proportion to the decrease in (real space) volume occupied. The same would apply for any (non-relativistic) initial energy so the overall effect for any distribution would be to increase all energies, and consequently the temperature by the same factor as the volume reduction.

$$T \propto V^{-1}.$$

Another way of looking at this result is to observe that, just as in the one-spatial-dimension case, the temperature change is proportion to C^2, the square of the linear scale.

To proceed to six-dimensional space of three spatial and three velocity dimensions, we simply combine the two processes.

Let us apply a magnetic field along the axis of our original vessel with the piston, now treated as a three-dimensional object, and consider what happens if we compress the volume with our original pistons by a compression ratio C and simultaneously reduce the radius of gyration of charged particles within it by the same factor, as indicated in figure 1.31. The real-space volume will reduce by C^3 while the volume of velocity space will increase by the same amount, the net effect being to keep velocity-space density constant. All particle energies will, as the result of one or the other, or both of these influences, be enhanced by C^2. The effect on energy will not be larger because none of the particles will be able to experience the full effects of compression by the pistons and orbit shrinkage by the magnetic field, since the former applies to motion parallel to the axis, and the latter to motion perpendicular to the axis. What we have then is

$$E_2 = C^2 E_1$$

and, since the volume has changed by a factor of C^3,

$$T \propto V^{-2/3}.$$

All three results of this exercise are anticipated in our experience of the behaviour of gases summarised in the gas laws [16]. One of these states that if a gas is compressed, without gaining heat from or losing it to the surroundings (other than through exchanging energy with the closing or expanding walls), a process known as adiabatic compression, temperature is firmly linked to the volume V by

$$T \propto V^{1-\gamma}.$$

The exponent γ depends on the number of dimensions the particle making up the gas have to move in, i.e. on the number of degrees of freedom N, thus:

$$\gamma = 1 + 2/N.$$

So for one degree of freedom, $\gamma = 3$; for two degrees of freedom $\gamma = 2$; and for three degrees of freedom $\gamma = 5/3$, which we have discovered as direct consequences of Liouville's theorem.

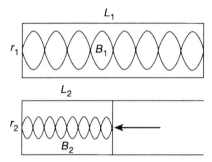

Figure 1.31. Compression in three dimensions of real space. The effects of the piston of figure 1.27 and the increasing magnetic field of figure 1.30 are combined. It is shown in the text that, again, the reduced real-space volume is exactly compensated by an enhanced volume in velocity space.

1.5.2.4 *Entry into a potential well*

Figure 1.32 demonstrates what happens to an energy distribution if charged particles are accelerated into a potential well. The effect is examined here, for simplicity, in one spatial dimension, the full three-spatial-dimensional case being reserved for Chapter 3. All we need do is note that all energies will increase by the same amount, namely the depth of the potential well in volts multiplied by the number of units of electronic charge carried by the particles, and that velocity-space densities will be conserved. The distribution inside the well is then just the distribution outside shifted bodily towards higher energies, as shown. If the particles found inside all stem from outside, there will, naturally, be none inside with energy below the amount by which energies increase. This results in there being a sharp peak in the distribution. The maximum velocity-space density is the same inside as outside.

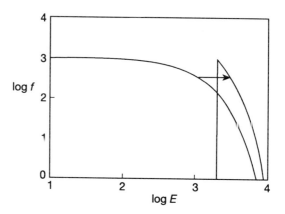

Figure 1.32. The effect of an attracting potential well on a Maxwellian. All particles experience the same energy increment (equal to the product of the depth of the well and the particle's charge). The resulting distribution has a cut-off at the common energy increment, creating a sharp peak at this energy. The high-energy tail has the same shape (on a log–log plot) as the initial tail—an observation which may be tested by tracing and superposition.

Figure 1.33 is the equivalent of figure 1.32 for particles climbing a potential hill. Here they all expend the same quantity of energy to lead to a distribution inside formed by a uniform shift towards lower energy. The maximum velocity-space density has now been reduced because the low-energy particles of which it was composed are unable to overcome the potential barrier opposing their entry.

Both types of distribution just discussed are created in the vicinity of potential non-uniformities created in or arising in assemblies of charged particles. A positive potential will attract electrons and give them a distribution similar to that of figure 1.32, and will, at the same time, repel positive ions to give them a distribution similar to that of figure 1.33. Since these happenings increase the number density of electrons and decrease the number density of ions, a positive deviation from uniformity tends to be cancelled out, or shielded by both effects. Negative disturbances are likewise shielded by accumulations of ions and reduced number densities of electrons. The result is that any disturbance to potential in an assembly of charged particles reduces exponentially with distance from the source of the disturbance. The attenuation length is known as the Debye length λ_D (e.g. [21]). It increases with the speed or temperature of the shielding particles, and decreases with their number density. For a neutral assembly of singly charged particles, having Maxwellian distributions with temperature T,

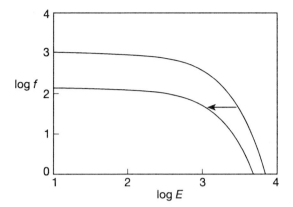

Figure 1.33. The effect of a potential barrier on a Maxwellian. All particles experience the same energy reduction. The final distribution is simply an attenuated version of the original—another observation that may be tested by tracing and superposition.

and electron and ion number densities n:

$$\lambda_D = 6.90\sqrt{T/n}.$$

A recent analysis [22] shows that for kappa distributions characterised by κ and E_κ

$$\lambda_\kappa = 6.07 \times 10^3 \sqrt{\kappa E_\kappa/(\kappa - 0.5)n} \qquad \kappa > 0.5$$

where E_κ is measured in eV.

We see, therefore, that Debye length, which is an important measure of a particle assembly, and which we shall employ as a yardstick in Chapter 2, is a natural consequence of velocity distribution modification caused by acceleration and retardation.

1.5.2.5 Constraints

The consequences of continuity enshrined in Liouville's theorem severely constrain the ways a distribution can be changed by dynamical forces. The energy at which a given velocity-space density or phase-space density occurs may change, but the corresponding densities themselves will be conserved. In figure 1.34 we see that this means that, under such forces, a distribution can be stretched or compressed in energy but not raised or lowered. In particular, the maximum velocity-space density remains a 'fingerprint' of the source through thick and thin. Just as a fingerprint can be used to identify an individual, maximum velocity-space densities can be used to identify, or narrow down, possible origins.

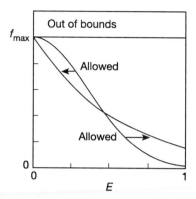

Figure 1.34. Constraints imposed by Liouville's theorem on the deformation of a velocity-space distribution. Deformation, by dynamical forces, is possible only through changes made at constant velocity-space density. Velocity-space densities in the modified distribution cannot exceed the maximum found in the initial distribution.

1.5.3   Loopholes in phase space?

By compressing particles into a smaller volume we have been unable to change velocity-space density (or phase-space density). We have, thus, seen something of the power of Liouville's theorem, and confirmed its compatibility with the gas laws for adiabatic compression. We shall see, however, when we come to measurements of electron velocity distribution in space that Nature does manage to produce intermediate velocity-space densities and even to create peaks and valleys. We must therefore persevere in our quest to find our equivalent of the alchemist's stone—a way to change velocity-space density.

One way would be to introduce fresh particles by 'creating' them *in situ* through ionisation or radioactive decay. Such irreversible injections will be seen to play a part in some situations, but there are many key areas where there are too few particles to ionise or to undergo decay, and too little ionising radiation to perform the task anyway.

Another process that seems, at least in its overall effect, to accomplish the feat is scattering. Scattering of directions can serve to reduce velocity-space density, or to appear to do so, when measurement is over a volume which may be only partially or temporarily occupied. Although the scattering agent may well be truly dynamical on the microscopic scale the average of many interactions may not be. A familiar example is the apparent grey scale of a photograph or newspaper picture which is an illusion created by a deployment of pure blacks and whites. Such a rearrangement of a stream of charged particles does not, though, modify velocity-space densities on the microscopic scale. Different parts of a finite thickness stream may be guided to different places, or the whole stream may be guided at different times to different places, but wherever and whenever there is a trajectory the velocity-space density will be conserved.

However, it is often the case that one's perception of events entails some measure of averaging. If this takes place over a region which is not fully irradiated, the measured intensity will be less than would be seen at a microscopic level. Scattering cannot, however, increase, even apparently, velocity-space densities above those in the original population.

Let us return to the piston illustrated in figure 1.27, but this time arrange, by some means (e.g. by conduction through the walls), for the energy gained by the particles under compression to be removed. Under these circumstances the reduction in volume will not be compensated by an increase in the volume of velocity space, so velocity-space density will increase. If the mechanism removes exactly all of the kinetic energy gained by the particles on colliding with the piston, velocity-space densities will increase in direct proportion to the compression ratio, while the velocities of the particles and the temperature of the distribution remain constant, as shown in figure 1.35. This constant-temperature compression is known as an isothermal compression.

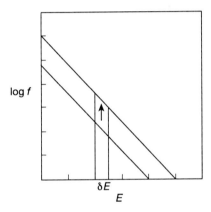

Figure 1.35. Isothermal compression of a Maxwellian. An assembly of particles under compression is maintained at a constant temperature by the removal of heat—an irreversible or non-dynamical process to which Liouville's theorem does not apply. Velocity-space density may thereby be increased uniformly at all energies.

Another non-dynamical effect that could, in principle, lead to changes in velocity-space density arises when the instantaneous acceleration depends on velocity. To illustrate this, let us return to our earlier analysis of deformation of rectangles in velocity-real space. This time, though, we shall arrange for the faster particles to experience a greater degree of acceleration than the slower ones. Figure 1.36 shows that in this case the 'volume' of the upper parallelogram is greater than that of the initial rectangle, making the velocity-space density fall. If we reverse the effect and accelerate the slower particles more than the faster ones, the volume shrinks and velocity-space density rises, as indicated by the lower parallelogram.

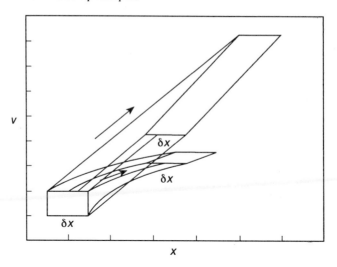

Figure 1.36. Demonstration in two dimensions that phase-space density can be changed by velocity-dependent acceleration. While the lengths of base and top of the initial rectangular two-dimensional 'volume' remain constant, the height of the new volume increases when acceleration increases with velocity (tall parallelogram), and decreases when acceleration decreases with velocity (short parallelogram). Velocity-space density falls in the former case and rises in the latter.

The new element which has allowed velocity-space density to change is velocity-dependent acceleration, which could result for example from dissipation, or energy loss to (or gain from), a third party. This leak (or accumulation) of energy constitutes a non-reversible, or non-dynamical element. Let us consider, as an example, a number of ball bearings being dropped from different heights into a tank of oil. They will enter the tank with a range of speeds, depending on the starting height. Within the oil they will continue to fall under the influence of gravity. They will also experience a viscous drag as they try to plough through the oil. The drag will be greater at higher speed and less at lower speed. The faster particles will therefore be dragged back closer and closer to the speeds of the slower particles until they all reach a speed where the force of gravity is balanced by drag, and a 'terminal velocity' is reached. The new element that has entered in this case is the viscous or frictional drag whose effect is to make the particles expend energy in heating the oil through friction. It is again, then, the element of dissipation which renders the net effect non-dynamical, thus permitting the increase of velocity-space density. The change in f can be shown to be under these circumstances [23]:

$$\mathrm{d}f = -f\,\mathrm{d}a/\mathrm{d}v.$$

Of crucial importance for our space plasma studies are the distortions that can be

shaped by non-dynamical processes circumventing Liouville's theorem. Figure
1.37 shows some of the possible effects on the energy distribution. The existence
of these possibilities is one of the main messages of this work.

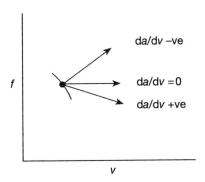

Figure 1.37. Transitions allowed for an element of a velocity-space distribution in
dynamical processes (where acceleration is independent of velocity $da/dv = 0$), and
non-dynamical processes (where acceleration depends on velocity $da/dv \neq 0$).

Liouville was careful to exclude specifically non-dynamical elements from
his theorem. We have not therefore discovered a flaw in the theorem, but we
have located an exclusion clause which we shall certainly exploit later.

1.6 Measurement and display

Before attempting to interpret electron distributions encountered in space and
to relate them to predictions emanating from theory it is important to realise
that we meet these distributions as measurements, and that all measurements are
subject to limitations, errors and uncertainty. Some comments, therefore, on the
techniques that are used, their particular advantages and inevitable shortcomings
might be found helpful.

1.6.1 Platform

In order to measure a particle velocity or energy distribution it is of course
necessary to transport a suitable detector into the relevant medium. In the
case of measurements in space, a rocket, satellite or other form of space
probe is required. Fundamental limitations immediately arise from unavoidable
perturbation of the medium by the vehicle. Space charge accumulated as the
result of the emission of photoelectrons under solar UV radiation can seriously
distort particle distributions near the spacecraft, preventing some from reaching
the detectors altogether, and giving rise to some that do not belong. Careful
design of the spacecraft, and, in some cases, control of electric potential by
emitting charged particles of the appropriate sign of charge, serve to minimise

the problem. Magnetic fields, stemming from magnetic materials that have to be included in the fabric of the spacecraft, especially where the magnetisation can change and thus become unknown, also cause a distortion of the true distribution, especially at low energies.   The motion of a spacecraft also introduces limitations. Since perceived particle velocities are actually velocities relative to the spacecraft, velocity distributions are, naturally, distorted by spacecraft motion, especially at very low velocities. A major problem arising at all energies from spacecraft motion is the ambiguity between spatial and temporal change. The problem can be alleviated to some extent by two or more vehicles travelling in tandem or from additional information such as photographs or other records from the ground. As a result of limitations of these kinds, the measurements we shall report and discuss rarely extend to electron kinetic energies below 10 eV, leaving the region below this, despite its sometimes crucial importance, largely unexplored.

1.6.2   Detectors

All detectors have a limited dynamic range.   However well designed and carefully built, they respond to stimuli other than those for which they are designed. Even in the complete absence of its nominal stimulus there will be a background due to impostors that have managed to get themselves into the reckoning. The practical result of this is that, in a medium in which low-energy particles predominate, as they do in most instances, measurements cannot be made over any arbitrarily wide range of energies, but are restricted to a range over which background counts are negligible. There is also the problem that, where particle fluxes are low, noise from the detector itself will begin to make a significant contribution. There is therefore usually a high-energy limit to the measurements achievable with a single detector. The high-energy limit can be raised by increasing the detector's aperture to allow through more of the wanted particles. However this has a repercussion at low energies where the flux into the detector will also increase and, if taken too far, swamp the detector so that it can no longer operate. The detector's inherent 'dynamic range', a figure of merit given by the maximum rate it can cope with divided by the noise rate, can usually be made to approach 10^5 or 10^6. While this might seem to be reasonably large, it does seriously limit the energy range that a single detector can encompass in a steep distribution, as a glance at the earlier Maxwellians, power laws and kappa distributions will confirm. While the use of several detectors offers an escape from this fundamental limitation, practical considerations of power, mass, space and telemetry eventually block this path. A compromise has to be made, and herein lies much of the skill of the experimenter.

1.6.2.1   The channel plate and channel electron multiplier

The versatile workhorse of charged-particle detection in space is the channel plate. The channel plate is an assembly of electron multipliers, developed initially for image intensification. This type of detector was first employed in space [24, 25] in single rather than the multiple form in which it originated and to which it has now generally reverted [27, 28]. The singular form, known as the channel electron multiplier, channel multiplier or channeltron, is simplicity itself. It consists, see figure 1.38, of a hollow tube made of a glass compound, open at one end and closed at the other. The inner surface is activated to become a material which conducts electricity, and from which electrons or other charged particle impacts can release one or more secondary electrons into the tube. The secondary electrons are accelerated along the tube by an electric field applied between a negative terminal, or cathode, at the open end of the tube and a positive terminal, or anode, at the closed end. Due partly to the tube being curved, in the form of a spiral or a helix, and partly to the secondary electrons having a component of velocity across the tube, the accelerated secondary electrons strike the surface some distance along. Like the primaries which started the process, they each stand a good chance of knocking out one or more secondaries. Repeated acceleration and emission snowballs into a cascade of ever-increasing numbers of secondaries making their way along the tube. The cloud finally arriving at the anode may consist of tens of millions of electrons. The deposition of these electrons causes a voltage pulse large enough to trigger an electronic counting circuit, so registering the arrival of the original charged particle. Other triggers of secondary emission at the cathode, notably x-rays, may also be detected [26, 27].

The curvature of the tube prevents any positive ions created within the device from being accelerated back to the cathode and initiating unwanted repeat cascades. The function is performed in the straight tubes of channel plates by constructing them in echelon.

The device as just described will do no more than just detect the arrival of charged particles of sufficient energy to release one or more secondaries from the cathode, rather as a photographic film responds to light. To gather information about the particle velocity distributions the detector needs something to help it discriminate between different prospective triggers, just as a camera needs a lens, an iris and, for some applications, a colour filter. The almost universal procedure is to place a filter in the form of an electrostatic analyser in front of the detector. One form of electrostatic analyser consists of a pair of curved plates, usually cylindrical, spherical or ogival, between which there is an electric field which guides charged particles of a chosen energy around the curve and into the detector [29]. Figure 1.39 illustrates the basic principles. The range of energies able to pass through the filter depends on the construction of the analyser, a long narrow gap being the most discriminating. The voltage between the plates is commonly varied through a series of steps or ramps in order to step or sweep

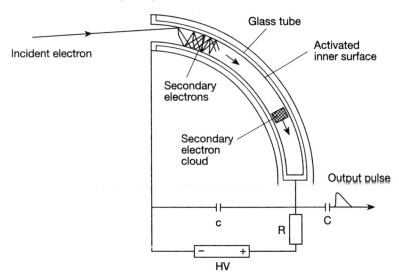

Figure 1.38. Principle of a channel electron multiplier. An incident electron (or other trigger such as an ion or photon) releases secondary electrons which, after acceleration in an electric field applied by means of a high-voltage (several kV) generator **HV**, release further secondary electrons. The sequence repeats to develop a cascade of 10^7 or more electrons to be detected as a voltage pulse across the stray capacity c of the anode. The pulse is fed, via a series capacitor **C**, to a counting circuit. The channel multiplier is re-charged through a series resistor **R** composed primarily of the electrical resistance of the channel multiplier itself.

through the energy range to be explored. The range of energies accepted by the device increases as the accepted energy is raised. This property of electrostatic analysers must, of course, be taken into account when deriving energy or velocity distributions from the measurements, as we shall see below.

A set of irises placed in front of the electrostatic analyser forms a (passive) collimator which allows the detector to select particles arriving from a narrow or wide field of view as required. The full picture is built up either by using a number of detectors looking in different directions or by scanning through the full scene as the spacecraft, or a platform mounted on it, spins around in space. More recently, the advent of microprocessors has made it possible for the different elements of channel plates to register simultaneously different arrival directions, or regions of velocity space. Channel multipliers or channel plates prefaced by electrostatic analysers are ideal for low-energy-particle sensors since, being open to space with no skin for particles to penetrate, they can work down to the lowest energies free from contamination and distortion introduced by the space platform itself.

It is possible to use magnetic analysers to filter incoming energies, and

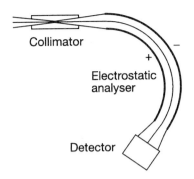

Collimator

Electrostatic
analyser

Detector

Figure 1.39. Principle of an electrostatic analyser used to select electrons or other charged particles from an incident flux. A collimator restricts and defines a range of directions. A voltage applied across a pair of cylindrical, spherical or ogival electrodes allows through to a detector only particles whose centrifical force is balanced by the radial electric field. The electrostatic analyser selects particles of a chosen sign of charge and chosen energy per unit of electronic charge. In order to explore a range of energies, the voltage applied across the electrodes may be stepped or swept through a range of values and/or a number of different analyser–detector combinations may be employed. Directional properties of the population may be explored using several combinations and/or by relying on the spin or other motion of the platform or space vehicle to sample different directions.

indeed this has been done on many occasions. However the method is rather less convenient, since magnetic fields are more difficult to contain, to produce and to change at will than electric fields. They also tend to require more of the precious and limited resources of power and mass than their electrostatic counterparts.

1.6.2.2 The solid-state detector

A detector in common use for electrons, and other particles with enough energy to penetrate its protective skin, is the solid-state detector [27]. This consists of a slab of semiconducting material with an electrode on each surface in the form of a disc capacitor. Charged particles that are able to penetrate into the body of the detector produce ionisation, the electrons of which are accelerated towards the anode producing a cascade of further ionisation, and depositing at the anode a space charge substantial enough to be detected as an electrical pulse. The process is very similar in effect to that occurring within a channel electron multiplier, except that it takes place through ionisation rather than secondary emission, and that the accelerating fields are adjusted so that the cascade does not develop to the maximum possible extent but remains proportional to the initial, triggering, quantity of ionisation. Since the initial quantity of ionisation is in turn proportional, at least over a limited range of energies, to the energy lost by a charged particle brought to rest in the medium, the output pulse is approximately proportional to the energy of the incident charged particle. Since ionisation can also be produced in the semiconductor through the photoelectric effect, the detector must be screened from solar and other UV by a thin surface membrane. This protective membrane is made thick enough to avoid the danger of holes which would allow some UV through and thus create a background noise, and thin enough so that electrons, or other particles, do not lose too much energy before reaching the sensitive volume. In practice, the thickness required limits the lowest detectable electron energy usually to around several tens of eV. An upper limit is set by the need to bring an electron to rest in the detector, in order to measure its full energy. Energies can be registered well into the MeV range.

Since a solid-state detector will respond to any particle (or other source of ionisation within its volume), it is necessary to place a filter in front of the detector. The types of filter used are of many forms, the simplest of which is a magnetic field that is weak enough to allow the relatively heavy and harder-to-deflect positive ions to pass through unhindered, but strong enough to sweep away the relatively light, more easily deflected electrons. Comparison with a control detector without such a magnetic 'broom' allows the entry rates of electrons and ions to be determined separately.

1.6.2.3   *Geiger and proportional counters*

The early stalwart of particle exploration in space, the Geiger counter [30] (in which an ionisation cascade develops in a low-pressure gas to the limit imposed by saturation), and its close relative, the proportional counter [26, 27, 30] (in which the discharge-tube cascade remains proportional to the initial degree of ionisation), have been largely replaced by the smaller, lighter and more robust channel plate and solid-state detector.

1.6.2.4   *The scintillator*

Another relatively early sensor, which remains in use for measurements at high energies, is the scintillator [26, 27]. This consists of a slab of translucent, luminescent material which, on being bombarded by charged particles, generates light which travels through the clear material to release photo-electrons from the cathode of an electron multiplier. The electron multiplier consists of a vacuum tube containing a photo-sensitive cathode and a series of secondary-emitting louvres, or dynodes, between which a secondary-emission cascade is caused to develop by an electric field applied between these elements. The principle is the same as that employed in the channel electron multiplier, with the continuous secondary-emitting surface of the latter being replaced in the photo-multiplier by a set of discrete surfaces.

1.6.3 Derivation of velocity distribution from measurements

How do we derive a velocity or energy distribution from the count rate of a detector or set of detectors? It will be clear from the preceding discussion that a detector cannot sample a distribution function in the way a botanist might count the number of plant species in a given area. A detector has to lie in wait, passively, for the different species, or different energies, to arrive and be counted over a specified time. To derive an underlying distribution from its measurements is in some sense equivalent to working out the population of insects in a garden from counting the numbers caught on a spider's web. A particle detector's basic currency is not numbers, but numbers per unit time or count rate.

Imagine that we have a detector with an aperture of area A, and with a collimator that allows particles to be accepted only if they arrive at a small angle to the axis of the detector, the solid angle of acceptance being Ω. Let the accepted particles lie in a small range of energy dE about a mean accepted energy E. The real-space volume filled by particles accepted in time interval δt is, at non-relativistic speeds, as the left-hand side of figure 1.40 shows,

$$V = Av\,\delta t.$$

The accepted volume in velocity space is seen from the right-hand side of the figure to be

$$V = v^2\Omega\,\delta v.$$

The combined six-dimensional volume, which may be thought of as the overall aperture for the measurement, is the product of these, i.e.

$$\mathcal{T} = A\Omega v^3\,\delta v\,\delta t$$
$$= 2A\Omega E\,\delta E\,\delta t/m^2.$$

$\mathcal{T}$ will have been determined by calibration and/or evaluation [29]. The number of counts recorded in an interval of time δt is given by the above product multiplied by the velocity-space density $f(E)$ at energy E in the sampled distribution, i.e.
$$N(E) = 2f(E)A\Omega E\,\delta E\,\delta t/m^2.$$

The count rate, $N(E)/\delta t$, is then

$$c(E) = 2f(E)A\Omega E\,\delta E/m^2.$$

In order to obtain velocity-space density from count rate, which is the usual task, we just rearrange the equations above to obtain, for electrons

$$f_e(E) = 1.62 \times 10^{-5}c(E)/A\Omega E\,\delta E \quad \text{km}^{-6}\,\text{s}^3$$

where E and δE are in eV.

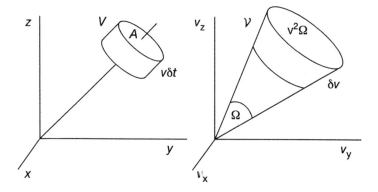

Figure 1.40. Volumes of real space V and velocity space $\mathcal{V}$ encompassed in a detector sample. A is the effective area, δt is the duration of the sample, v is the central velocity, Ω is the acceptance solid angle and δv the range of accepted velocities. These combine with the count rate c to yield velocity-space densities.

If the analyser is an electrostatic one, the bandwidth δE will, at non-relativistic energies, be proportional to E, making

$$f(E) \propto c(E)/E^2.$$

Due to the form of the relationship between energy and velocity, velocity-space density loses its usefulness in the relativistic regime, so we shall not concern ourselves with a general expression for this quantity.

Phase-space density is given in the non-relativistic regime simply by $f_e(E)$ divided by the (electron) mass cubed, i.e.

$$F_e(E)_{NR} = 2.15 \times 10^4 c(E)/A\Omega E\,\delta E \quad (zJ\ s)^{-3}$$

(with E and δE continuing to be expressed in eV). A general expression for electron phase-space density, holding for any velocity, is

$$F_e(E) = 2.20 \times 10^{10} c(E)/A\Omega(E^2 + 2E_0 E)\,\delta E \quad (zJ\ s)^{-3}.$$

In the ultra-relativistic regime this simplifies to

$$F_e(E)_{UR} = 2.20 \times 10^{10} c(E)/A\Omega E^2\,\delta E \quad (zJ\ s)^{-3}.$$

The differential intensity, a frequently used measure—the number of particles crossing unit area, per unit solid angle, per unit energy—is given (in basic units) by

$$j = c(E)/A\Omega\,\delta E\,\delta t$$
$$= p^2 F.$$

For electrons

$$j_e(E) = 10^3 c(E)/A\Omega\,\delta E \ \ \mathrm{m}^{-2}\ \mathrm{s}^{-1}\ \mathrm{sr}^{-1}\ \mathrm{keV}^{-1}.$$

This holds for all energies (in eV).

1.6.4 Methods of presentation

How are distribution functions, once measured, to be represented? Since velocity- and phase-space densities generally depend not only on energy and direction but also on position and time, there is no convenient way of depicting all aspects at once. Recourse is, therefore, made to the presentation aspects crucial to the study in hand. As a result, many types of display are employed. A saving grace, especially for electrons in space, is that the distributions are quite often approximately symmetric about the local magnetic field direction, thus giving a cylindrical symmetry or gyrotropy. There is sometimes complete spherical symmetry, or isotropy.

A widely used type of presentation is that of the energy distribution sampled at a given place and time with specified direction of travel. This may be of phase-space density, velocity-space density, intensity or count rate. Figure 1.41, the first of a set of different presentations of the same hypothetical example, shows an energy distribution, with velocity-space density as the variable. 'Perpendicular' particles ($\perp$) with pitch angles close to 90° have a monotonic distribution, while the 'parallel' particles ($\parallel$) show a slight peak. Figure 1.42 has the same information plotted as an energy spectrum of intensity against energy. The example distribution is now seen to have a well defined peak in intensity in the parallel direction.

A measure of the angular dependence through the full range of pitch angles may be gained from figures such as figure 1.43, which in this case gives the angular dependence for three representative energies. Energy and angular dependence may be combined in 'surface' plots such as figure 1.44 which present a 'view' of velocity-space density against both energy and pitch angle. These are particularly useful for gaining an overall impression [31].

Another way of combining energy and angular dependence is through the contour plot. In figure 1.45 contours of velocity-space density are traced on a map whose latitude and longitude are the parallel and perpendicular components of velocity. A distinction is made here between velocity components parallel to the magnetic field $v_\parallel$ and oppositely directed, or anti-parallel, components $v_\#$. Sometimes, as here, the plot is reflected about the axis of zero perpendicular velocity ($v_\perp = 0$) so that the lower half mirrors the upper half. This serves two useful purposes. One is to make the contours easier to assimilate (it is much easier to judge, for example, a true circle than a true semi-circle). Another is to permit experimental measurements of the same perpendicular velocity taken at different azimuths to be shown separately, thus affording a check on the gyrotropy implied by the use of this method of presentation.

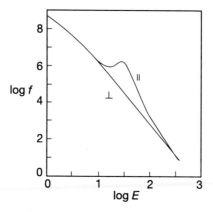

Figure 1.41. The energy distribution, consisting of velocity-space density f as a function of energy, is one of many methods of depicting an electron, or other particle population applying at a given place and time. The different profiles for electrons travelling in different directions—here approximately parallel ($\parallel$) and perpendicular ($\perp$) to a reference direction such as the magnetic field—indicate that the distribution is anisotropic.

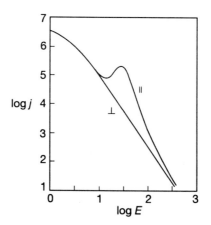

Figure 1.42. The energy spectrum, consisting of intensity j as a function of energy, is plotted for the same electron population as that represented in figure 1.41. The greater emphasis given to higher energies in this method of presentation is helpful in accentuating features such as the slight peak of the energy distribution.

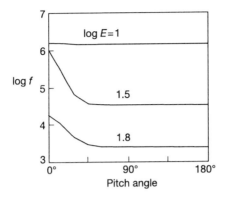

Figure 1.43. Pitch-angle distributions. Details of any anisotropy may be revealed by plotting f as a function of pitch angle α for a number of representative energies. The peak of our example distribution is seen to be confined to pitch angles below about 15°.

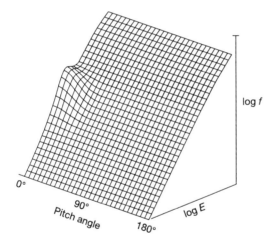

Figure 1.44. A surface plot of f as a function of energy and pitch angle provides an overall view of a particle population.

Simple contours may be used as here, the bands between the contours may be colour coded as in standard contour maps or there may be a continuous graduation of colour or grey scale. While these are all designed to improve assimilation, problems do arise in that colour perception varies between individuals and it is often very difficult to recognise different levels of grey well enough to match them against the code given at the side. These are fundamental physiological problems stemming from the fact that our assessment of colour and brightness varies with the background against which the sample is seen.

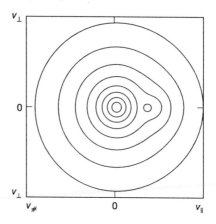

Figure 1.45. A contour plot of sample values of f as a function of velocities parallel and perpendicular to the magnetic field provides a convenient all-round assessment of a distribution, especially for gyrotropic ones. Velocities parallel and anti-parallel to the magnetic field are distinguished by the subscripts $\parallel$ and #. Contours are spaced at order-of-magnitude intervals, the highest at the centre, and are reflected about zero perpendicular velocity.

We shall return to this point later in connection with perception of structure and patterns within the aurora. Grey scales are sometimes used to avoid the expense of colour reproduction. They tend to be disappointing and less informative overall, though careful tailoring of the scale to highlight key features can regain some of the lost ground.

Numbered contours avoid these problems, but, in a complex case are much more difficult to assess at a glance, and the annotations tend to obscure important detail. On some occasions, a set of contours derived from measurements is overlaid with contours predicted by a particular theory with a view to testing the applicability of the theory. A question that immediately arises is that of deciding how good or how bad the fit is. To my knowledge this vital question has not been considered in a quantitative way, if at all. This will need to be taken in hand at some stage. One suggestion would be to note whether or not the experimental contours cross the proposed ones. Good agreement is manifested by little or no crossing; poor agreement by crossing at large angles.

A particularly striking form of presentation is that of a surface plot of velocity-space density as a function of parallel and perpendicular velocity. A surface plot for the present example is shown in figure 1.46.

Dependence of the distribution on time or position is usually displayed by plotting the chosen variable at representative energies at regular intervals of time, as for example in figure 1.47. Cinematography has some particular advantages, too. A sequence of any of the previous forms of display viewed as moving pictures exploit our instinctive sensitivity to movement and can reveal subtle

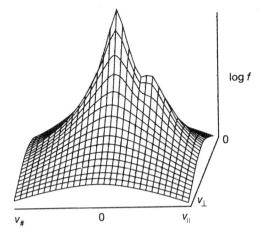

Figure 1.46. The mountain-like form of a surface plot of f as a function of parallel and perpendicular velocity is often both striking and revealing.

changes or trends which are very difficult, if not impossible, to detect from a sequence of still pictures. There remains, though, the challenge of capturing any fleeting progression observed by such means.

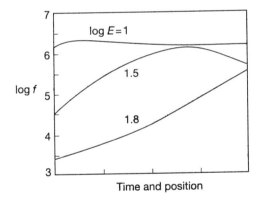

Figure 1.47. Time series of velocity-space density or other variables at several representative energies provide a measure of the way a distribution varies with time and position combined.

A form of presentation that has been found particularly helpful in space research is the spectrogram. The energy–time spectrograms shown in the three parts of plate 1 all have energy as ordinate and time as abscissa. The colour-coded scales are of velocity-space density in plate 1(a), intensity in plate 1(b), and count rate in plate 1(c). These figures are, in essence, horizontally stacked time series of energy distributions as seen from a detector or detectors mounted on a moving vehicle. The 'time' axis therefore represents a combination of temporal and spatial components. In recognition of this, it is common to give several scales to the time axis to show how other variables, e.g. satellite altitude, latitude and longitude vary at the same time. When the time resolution is high enough to sweep through several energy scans in the time that the detector sweeps through a range of directions, fine structure in the spectrogram reveals any directional dependence within the velocity distribution. Alternatively, measurements taken at different arrival directions are displayed in a family of spectrograms.

Interpretation of such figures demands some care, as the different appearance of the three forms of display of the same data makes very clear. This demonstration shows what would be observed if the peak in velocity-space density, seen in figure 1.41 emerged steadily from an original distribution without a peak, and progressed at constant height towards higher energy. The earlier spot samples correspond to the mid point of this progression. While velocity-space density is in many senses the most physically meaningful reference quantity, the wide dynamic range to be accommodated (see the appended scale) often severely limits the resolution. Intensity, with its reduced dynamic range, is sometimes chosen as a useful compromise between velocity-space density and count rate (also often preferred), which gives the highest resolution and at the same time indicates which particles are conveying most energy.

1.7 Conclusion

We are now equipped with most of the general tools needed to explore electron acceleration in space (others will be introduced when needed), and have a fully quantitative understanding of how to use them.

Chapter 2

The arena

In this chapter we survey the arena within which the electron acceleration to be explored in the following chapters takes place. We look at its nature, its configuration and its electron content. The medium is an unfamiliar one and yet at the same time the one that is most common in the universe. The arena is vast. It ranges through the Earth's outer environment, into interplanetary space, and on to the Sun and interstellar space. The electrons populating it span more than ten orders of magnitude in energy from the lowest currently measurable to the highest theoretically attainable. Phase-space densities encompass more than 40 orders of magnitude.

2.1   The nature of the medium

2.1.1   The plasma state

Number densities of particles throughout the arena are, by any usual standards, extremely low. A cubic metre of air at sea level normally contains $\sim 10^{25}$ molecules. At altitudes of 100 km, where our 'space' begins, the number density falls to around 10^{19} m^{-3}. By 500 km it is down to $\sim 10^{12}$ m^{-3}, a lower number density by far than can be achieved by any man-made vacuum pump. Most of the number densities we shall be considering in the Earth's outer environment and in interplanetary space are very much lower still at 10^7 m^{-3} and below. There is a sense, then, in which this arena can be considered an empty void. However, this would be to overlook the enormous volumes and the vast numbers involved and would be to misread the true situation. The electrons we shall be studying have energies that are too high for them to be retained in the most loosely bound positions in atoms, or for them to become trapped by protons or positive ions of heavier elements to complete the sets required to make atoms. The element with the highest ionisation energy is helium. At just under 25 eV this lies almost at the lower limit of the range of energies open to exploration. The medium is, thus, an assembly of positive ions (principally protons) and electrons, with an important admixture of ions of other elements. Collisions, in the form of interactions between the electric fields of individual particles (ions with ions, electrons with electrons or with ions) are so rare as to play only a minor role in what happens, except at the lowest terrestrial and solar altitudes where 'space' merges into an atmosphere. As a result, quantities familiar to us on Earth such as viscosity and electrical resistance are negligible in space. Interactions mimicking these properties do take place, though, between the constituent particles and the collective fields of others.

This electrified state of matter—the most common state of matter in the universe—is known as a plasma (there being no scientific connection with its namesake in medicine). Plasmas represent, after solids, liquids and gases, the fourth and hottest state of matter. In such a medium every particle, through the electric field of its charge and the magnetic field associated with its motion, has an influence on its neighbours. The behaviour to be expected of a plasma is, therefore, considerably more complicated than that of a gas of predominantly neutral particles which interact by colliding one to one. The relatively long range and collective interactions which take place in a plasma make the medium a lively and complex one, even in the rarefied conditions of space. It might at first seem to those versed in the plasmas created in laboratory devices, employed for example in the drive towards controlled nuclear fusion, where number densities of electrons and ions may be 10^{19} m^{-3} or more, that space plasmas might be too diffuse to be thought of as comparable in any way. This would again be to misread the situation. In a very real sense—that which governs the physical properties and behaviour of the medium—the plasmas of space may be considered to be very much more dense than their laboratory counterparts, and to be the most perfect examples of the plasma state available for us to explore. To substantiate this claim, we need to become more quantitative, and examine the dividing line between a plasma and its near relative, a partially ionised gas consisting of a mixture of charged and neutral particles. There are three main criteria [32] to be met for plasma behaviour to dominate gas behaviour.

The first is that a plasma should have room to operate. It must occupy a large enough volume for interference or contamination from its boundaries to play no significant role. A useful yardstick for measuring the space a plasma needs is the distance it takes the plasma to neutralise or screen any concentration of charge that happens to build up. It does this by surrounding and cancelling out the excess with particles of the opposite sign of charge. The screening distance is the Debye length mentioned in Chapter 1. The first criterion, then, crystallises into a requirement that the linear dimensions of the volume occupied by the plasma are very much greater than the screening distance. The Debye length, λ_D, in a plasma where the electrons and ions have the same mean energy $\langle E \rangle$ and the same number density n is,

$$\lambda_D = 6.07 \times 10^3 \sqrt{\langle E \rangle / n} \text{ m.}$$

In the solar wind, where $\langle E \rangle$ is around 15 eV and n at the Earth's orbit is typically 10^7 m^{-3}, λ_D is typically about 7 m. In the plasma sheet, where $\langle E \rangle$ is much higher at 1500 eV and n is typically 10^6 m^{-3}, λ_D is about 200 m. The regions occupied by these plasma are on vast scales compared to these values, so the first criterion is well satisfied in these cases, and indeed in all media we shall be concerned with.

The second criterion is that the plasma keeps itself almost exactly electrically neutral. This amounts to a requirement for very large numbers of particles to be involved in the screening process. A convenient test is to check that the number of particles contained in a sphere of radius equal to the Debye length is very large. From the numbers above, we can see that for the solar wind this number, N_D is approximately 10^{10}, and for the plasma sheet N_D is approximately 10^{12}. These are by any standards very large numbers, much higher than any that can be achieved in laboratory plasmas, so the second criterion is, like the first, very well satisfied.

A third criterion is that collective interactions, rather than individual ones, determine what happens. A convenient test for this state of affairs is to see whether there is time for a particle to complete many oscillations characteristic of the plasma before it encounters another particle too closely. One of the most characteristic oscillations is the oscillation of a particle under the restoring force set up when it is displaced and the neutrality is locally upset. This plasma oscillation depends on the electron (and ion) number density, n, of the plasma. The plasma period for an electron is

$$\tau_{pe} = 0.11/\sqrt{n} \text{ s.}$$

For protons the plasma period is 42.8 times longer by virtue of the square root of their greater mass. In the solar wind, the plasma periods for electrons and protons are typically 30 μs and 1.3 ms, respectively. In the plasma sheet, the electron and proton plasma periods are 50 μs and 3 ms. Before allowing these to set the standard, we need to note the times required for another fundamental plasma process—gyration in a magnetic field.

The gyro-period for a given type of particle depends, as we saw, on the mass of the particle, on its charge and on the magnetic field strength, B. An energy dependence arises only at relativistic energies (Chapter 1). The gyro- or cyclotron period for an electron is

$$\tau_c = 3.57 \times 10^{-2} \Gamma / B \text{ s}$$

where B is in nT. For a proton the gyro-period is 1834 times longer, due to the greater mass. In the solar wind, where B is typically 10 nT, the gyro-periods for electrons and protons are 4 ms and 7 s, respectively. In the plasma sheet, where B is typically 1 nT, the gyro periods are 40 ms and 1 min. With these plasma and gyro-periods we now have a set of standards against which to measure the times between various types of inter-particle collisions.

The mean time taken for an electron to pass by chance close enough to a proton to be significantly deflected, or perhaps captured to complete an atom of hydrogen, may be estimated as the mean time to pass within a distance of the order of the radius of such an atom. Taking this to be one Bohr radius $(5.29 \times 10^{-11}$ m), the mean time between such collisions for non-relativistic electrons is, following [16],

$$\tau_{coll} = 3 \times 10^{20} nv \text{ s.}$$

In the near-Earth solar wind, the collision time for an electron of 10 eV $(v = 1.87 \times 10^6$ m s$^{-1})$ is around six months. In the plasma sheet, where the number densities of charged particles are typically ten times lower, the time for collisions between charged particles is measured in years. Collisions of this type will clearly not interfere with plasma oscillations and gyrations. In the plasma sheet there are substantial numbers of hydrogen atoms, forming a 'geo-corona' [33]. In fact, here, these neutrals outnumber charged particles, typically by about 100:1. Collision times are reduced, accordingly, to 2 weeks. This is no limitation either.

Collisions of electrons with electrons where the target is much smaller (the classical radius being 2.82×10^{-15} m) are negligible by comparison. Proton–proton collisions are also negligible because, although they make larger missiles and targets, their speeds are very much lower.

Electrons of the regions in which we shall be interested are, thus, able to complete, in very round numbers, $\sim 10^{12}$ plasma periods and $\sim 10^9$ gyrations without interruption. Even plasma sheet protons can expect to complete a billion oscillations and a million gyrations before encountering, and exchanging charge with, hydrogen atoms. The third criterion is, therefore, like the others, met rather handsomely for electrons and reasonably well for protons (though confirmation is necessary in some specific cases).

The medium we shall be investigating can, therefore, with some justice, be thought of not, as might first have been supposed, as one which is too diffuse to support plasma behaviour, but as a giant dense plasma which comes closer to the ideal than a plasma produced in a laboratory ever could. The enormous sizes of space plasmas offer great advantages when it comes to measurement, and therefore lend themselves to very close and detailed investigation.

2.1.2 Magnetic lines of force

One of the key properties of a plasma is the magnetic field it contains. It is often convenient to consider this in terms of magnetic lines of force. A magnetic line of force is a concept invented by Faraday to demonstrate force at a distance [35]. It is a locus of points that would be followed by moving a miniature compass needle along the direction in which it was pointing. It is important to note, though, that if the field being traced by this method has small-scale bends and turns, the 'line' traced out will depend on the size of the compass. Similarly, if the field varies with time, the 'line' traced out will depend on the time of the operation. While there may be, at any given instant of time, a hypothetical line of force that would be traced by an infinitely small compass travelling at infinite speed, this is not really a concept that has physical meaning or application. What is often of concern in space plasma studies is the trajectory that would be followed by a particular charged particle following its approximately helical path around and along the direction of the field. Because actual trajectories vary with particle properties, such 'field lines' will be different for particles of different species, different energies, different initial directions of motion and different times. The treatment of field lines as distinct entities or objects which can move from one place to another, and can be stretched, cut, and joined as though they were made of rope, must, I feel, be treated with great caution, i.e. the approach needs to be justified by conventional means in each and every case.

2.2 Interplanetary space

Interplanetary space [36] is a logical starting point for the mapping of the arena because its content, the stream of charged particles known as the solar wind, has a dominating influence on the outer environments of the Earth and, indeed, the other planets of the solar system. The solar wind is the the Sun's expanding outer atmosphere, or corona. It consists of electrons and protons, with an admixture of helium nuclei and other solar constituents.

2.2.1 Protons and other ions

While protons and other ions are not the subject of this work, they are highly instrumental in shaping the environment for our electron studies. It will be helpful, therefore, to sketch in the general picture. The protons and other ions of the solar wind form well defined beams in velocity space, the beams being directed radially, or almost radially, from the Sun. To a first approximation, the ion beams are drifting Maxwellians, in which the outward drift velocity, which has a mean value of around 430 km s^{-1} (one million mph), is more than two orders of magnitude greater than the random, or thermal, speed. The number density and drift velocity vary widely from time to time, and from place to place, in the ranges 3×10^6 to 2×10^7 m^{-3} and 200 to 1200 km s^{-1} respectively, e.g. [34].

The mechanisms giving rise to the solar wind and its varied properties are still the subject of much discussion. While escape of high-speed particles from the gravitational field of the Sun plays a role, it is not the whole story. It has very recently been confirmed that solar wind emerging from higher latitude regions of the corona generally flows faster than that from low latitudes [37]. The high-latitude wind is also generally less dense. The latitude referred to here is the latitude with respect to the magnetic axis which, when the Sun is at the most active phase of its 11-year cycle, is tilted quite markedly (up to 30°) with respect to the rotation, or spin axis. The combination of tilt and spin produces a band of latitudes where wind speed and density vary during each solar rotation of approximately 27 days. The band of latitudes includes the ecliptic plane where the planets lie, and where until relatively recently all direct measurements were confined.

2.2.2 Magnetic field

The energy per unit volume, or energy density of the ions is, at all points further than several solar radii from the Sun, considerably greater than that of the ingrained magnetic field. The regime is, therefore, one in which the ions dominate, and control how the magnetic field extends into space. If the Sun did not spin on its axis, the magnetic field would follow the flow lines of the ions, and extend radially from the Sun. But the Sun does spin. It spins on its north–south axis, in the same sense as the Earth, with a period, averaged over all latitudes, of approximately 27 days. As a consequence of this, the locus of solar-wind particles emanating from a given part of the corona is an Archimedian spiral, whose radius increases in proportion to the rotation angle (figure 2.1). The effect is analogous to that which makes the locus of a stream of water droplets issuing from a rotating garden hose or sprinkler trace out its familiar curve. Due to the high electrical conductivity of the solar wind, any attempt to change the magnetic flux within any element of the plasma is thwarted by currents induced within the element. If, as is usually assumed, the magnetic field at the point of release is precisely radial, all plasma elements streaming from the same point will begin their career into space magnetically connected, or magnetically conjugate and will remain so at all times. At distances r more than a few solar radii the deviation θ of the magnetic field from the radial direction is given, when the solar-wind speed is v_{sw} by

$$\tan \theta = 2.7 \times 10^{-6} \, r/v_{sw} \, .$$

This expression shows that at a typical solar-wind speed of 430 km s^{-1}, the inclination at the Earth (where $r = 1.5 \times 10^{11}$ m) is 43°. At Mars ($r = 2.3 \times 10^{11}$ m) the typical inclination is 55°, and at Jupiter ($r = 7.8 \times 10^{11}$ m), to take another example, it becomes 80°. The angle gradually, under these influences alone, approaches 90° at very large distances. Other factors come eventually into play, though, as we shall report.

It follows that even slight departures ($< 1°$) from radial flow, or radial field close to the Sun project into gross departures from nominal, leading even to reversals in the sense of the spiral.

Typical solar-wind ions take approximately four days to reach the Earth. In this interval, the Sun rotates approximately 4/27ths of a full rotation, i.e. 54°. This means that if a line of force could be traced back along the garden-hose spiral from the Earth to the Sun, the tracker would arrive at a point close to the western edge or limb of the Sun, as seen from the Earth. While changing conditions, especially of flow speed, make this magnetically conjugate point somewhat variable and distinctly hazy, this straightforward geometrical

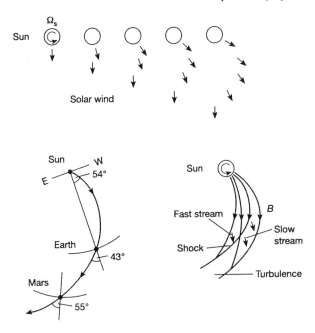

Figure 2.1. The solar wind and the interplanetary magnetic field. The top part of the figure shows how the locus of elements of solar wind issuing from a given location on the Sun, which rotates with angular velocity Ω_s, forms a spiral pattern in the ecliptic plane. Due to the high electrical conductivity of the plasma, a magnetic field that is radial at the point of emission is drawn out into a 'garden-hose' spiral, as illustrated in the lower parts of the figure. For a typical solar-wind speed of 430 km s^{-1} the interplanetary magnetic field is inclined to the radial direction by 43° at the Earth and 55° at Mars. Under these conditions both planets are magnetically conjugate to (generally different) regions of the Sun. As seen from the planets these are, in the case of the Earth towards the western edge, or limb, and in the case of Mars almost at the western limb. Variable solar-wind speeds lead (lower right) to gross departures from this picture, and to both shocks and turbulence in interplanetary space and the wider heliosphere (from [57]).

consideration serves well to account for the experimental fact that particles released in solar flares, finding it easier to follow helical paths about the field direction than to make their way across a field barrier by scattering from irregularities, are more readily able to reach the Earth from flares occurring near the western limb than from the centre or, still more, the eastern side of the Sun.

The magnetic field set up in interplanetary space is, in practice, rather more complicated than the 'garden-hose' picture given above especially when the Sun is variable or 'active'. One reason for this is that, just as with the Earth or any other magnetised body, field lines in different hemispheres have opposite radial components. The 'wobble' and polarity variations of the magnetic field configuration as the Sun rotates with its inclined magnetic axis causes the field at low latitudes to alternate in direction. This commonly occurs four times in a rotation period corresponding to there being four 'sectors'—two of outwardly directed field and two of inwardly directed field. Some sector patterns retain their general properties for many rotations. Temporal changes and spatial structure in number density and speed introduce great complexity. High-speed wind emitted from one region may overtake a more slowly moving patch emitted earlier in the same direction from another, causing a disturbance to solar wind and magnetic field in the region of overlap. There are similar effects when a slow region is unable to keep pace with a faster region ahead. The sector structure is, then, frequently accompanied by turbulent boundaries where particles emanating from different regions with different properties interact. Because of the low densities, interaction is unlikely to be between individual particles, but between particles and the wave fields set up collectively by others. Our earlier discussion of the general effects of such random exchanges of energy would lead us to expect (rightly) that the overall effect is one of broadening of velocity distributions, i.e. heating of the particles. The energy for the heating is naturally drawn from the drift motion, which is consequently reduced.

Major perturbations to the solar wind and consequently to the interplanetary magnetic field are caused by solar flares. Solar flares, which are sudden, short-lived and highly localised events originating, it seems, in the outer corona, release protons, other ions and electrons of relativistic energies into interplanetary space. As just discussed, the vanguard from flares near the west limb reaches the Earth in just a few minutes, only just after the light and x-rays from the flare itself. Others take much longer. They take so long in fact that the median time corresponds to a journey of many times the Sun–Earth distance, proclaiming a tortuous path through interplanetary space. Some flares and other active regions emit also a very much enhanced flux of solar wind. This stream (or coronal mass ejection) is of too high an energy density to follow the pre-existing 'garden-hose' field, but bursts through this to set up a new pattern.

Except where the solar wind meets obstacles to its progress such as planets or comets, it continues to flow outward with undiminished speed and geometrically reducing number density. Eventually, the geometrically reducing energy density obliges the solar wind to submit to the superior energy density of other stellar winds making up the interstellar medium. The very distant, and as yet unexplored boundary, where this hand-over of sovereignty occurs, is known as the heliopause. The volume of space under solar-wind jurisdiction is known as the heliosphere.

2.2.3   Electric field

A question that might arise in connection with figure 2.1 is that of how the ions of the solar wind can flow at an angle to the interplanetary magnetic field without being deflected. The fact that they themselves have set up the field, as we have just seen, does not itself amount to an explanation. We are still faced with the fact that protons whose gyro-radius in the interplanetary field is of the order of 10^6 m are found at and well beyond 10^{11} m from the Sun, still travelling more or less radially. They have clearly covered distances many orders of magnitude greater than their gyro-radius without significant deflection. The question becomes, on the face of it, even more puzzling when we note that gyro-radii are even shorter in the stronger fields closer to the Sun.

The answer lies in one of the most influential processes serving to shape space plasmas. The process relies on the existence of a key ingredient—space charge. Imagine what would happen if solar-wind particles were deflected by the magnetic field—ions one way and electrons the other, in accord with their opposite signs of charge. For simplicity we can consider the magnetic field to be at right angles to the radial flow, as it tends to become at large distances from the Sun. The charge separation effected by the deflection would give rise to an electric field tending to oppose and so reduce the deflection of subsequent particles. Later groups of particles would experience stronger and stronger restoring fields which they would themselves enhance until a balance was achieved between the magnetic force tending to deflect the particles and the electric force sending them back. Once this natural and automatic balance is achieved, there is no net force on subsequent particles. They are then able to proceed, force free, in straight lines.

The situation is analogous to charge separation in a dynamo. Consider a metal conductor moving through a magnetic field. At first, the particles free to move within the conductor do not move exactly in the direction of the conductor itself, but are deflected to concentrate on one of its flanks. Electric charge builds up until the force on the electrons exerted by the electric field so produced balances the deflection force due to the motion through the magnetic field.

The strength of the electric field which gives zero net force is that which makes the net Lorentz force zero. With v in m and B in nT we have, for v perpendicular to B,

$$\mathcal{E} = -10^{-9}vB \ \text{V m}^{-1}.$$

In the region of the Earth's orbit, B is typically around 5 nT, so, with a typical v_{sw} of 4.3×10^5 m s^{-1}, the electric field is typically around 2 mV m^{-1}.

The quantity and spatial arrangement of space charge needed to generate such an electric field has, as far as I am aware, not been evaluated. The difficulty is that it will depend on the full three-dimensional picture of the solar wind, which is as yet unknown. It will depend also on the degree of (Debye) screening of the charge by neighbouring particles.

Let us, however, make an estimate of the quantity of space charge required, in an attempt to see whether this represents a major or minor perturbation of the medium. We will assume for simplicity that the space charge producing this field is arranged in two slabs of thickness η lying parallel to the ecliptic plane at a spacing much less than their lateral dimensions. Let them both have the same number density n_q of excess charge, one positive and the other negative. In the absence of other factors, the electric field in the space between them would, by Gauss's law, have a strength

$$\mathcal{E} = 3.6 \times 10^{-8} \eta n_q \text{ V m}^{-1}.$$

Equating this to the 2 mV m^{-1} estimated above we obtain

$$n_q \sim 5 \times 10^4 / \eta \text{ m}^{-3}.$$

Unfortunately we do not know η and have no means of estimating it. However, we can note that even if the space charge is concentrated into very thin (on the interplanetary scale) slabs of $\eta = 1$ m, the ratio of n_q to n, the unperturbed electron and ion number density, would, for $n \sim 10^7$ m^{-3}, be only 1 in 200. Since η can in practice be expected to be much larger than this, it is clear that the perturbation represented by the space charge, though of immense importance to the electric fields and dynamics of the region, is likely to be very small indeed.

Note that the whole configuration of space charge has to be re-established each time there is any change in solar-wind velocity or in the magnitude or direction of the interplanetary field. It must also be maintained against the natural tendency of the medium to shield or neutralise it.

The existence of the space charge, and the electric field which stems from it has many important consequences in space. One of these is the phenomenon known as 'pick-up' [38]. Any charged particles that become deposited, by ionisation or other non-conservative process, in the solar wind, but which are not moving in sympathy with it will find the magnetic force and electric force out of balance, and will become accelerated by the electric field as a consequence. The net effect of this acceleration and the deflection by the magnetic field causes both electrons and ions to be 'picked up' and eventually subsumed into the solar wind. This process will be discussed further and in a number of other contexts subsequently.

2.2.4   Electrons

Electrons of the solar wind are very different and behave very differently to the ions. Energies extend over a broad range one to two orders of magnitude below that of the almost mono-energetic ions. Typical velocities are, despite this, very much greater. A consequence is that the interplanetary electric field is too weak to compensate for the magnetic deflection of electrons, so the electrons are forced to gyrate about the magnetic field. Any directional properties tend to be aligned with the magnetic field, making the distribution 'gyrotropic', and often almost isotropic.

Six examples of electron energy distributions obtained by the UKS spacecraft [39] are shown in figure 2.2(a). The distributions are noticeably curved and therefore not true power laws. They are better represented overall by kappa distributions with values of K ranging from K = 3.5 to 10 eV, and exponents ranging from −3.4 to −4.8. Figure 2.2(*b*), demonstrates this by yielding straight lines over the greater part of the distributions when the abscissa is log(E + K), each distribution having its appropriate K. Solar-wind electrons will be discussed as sources for other populations in Chapters 4, 5 and 6, and the origin of the electrons themselves is the subject of Chapter 9.

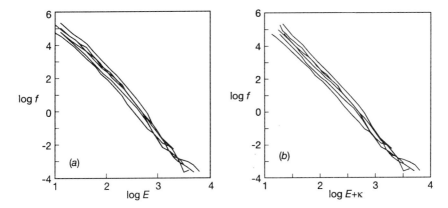

Figure 2.2. Energy distributions of solar-wind electrons encountered by the UKS spacecraft. Note in (*a*) the approximately power-law form, with a perceptible increase of slope toward higher energies. A closer fit is obtained with a kappa distribution, as indicated by the approximately straight lines in (*b*).

2.3   The solar wind in the vicinity of planets and other bodies

2.3.1   Magnetised bodies

We saw that the solar wind is able to establish conditions in interplanetary space that allow it to flow freely. When it encounters a planetary magnetic field, however, it meets its match. Although it can dominate the planetary field at great distances from the planet, it eventually reaches a stand-off point and has no option but to flow around the obstacle. A solar wind that is denser has a greater ability to generate induced currents, and thereby magnetic fields, to cancel out a planetary field. Higher wind speeds also enhance this capability. A combination of these attributes that serves as a useful measure of the bulldozing capability of the solar wind is, not surprisingly, the ram pressure. Ram pressure is the product of the number density and the square of the drift velocity. Since by far the greater part of the energy of the solar wind is contained in the drift of the protons, the ram pressure equates to twice the solar-wind energy density. Note that for any given state of the solar wind, the density falls off roughly as the square of distance from the Sun, so the outer planets are able to keep the solar wind at bay over more extensive regions of space than they would be able to do nearer the Sun.

Whenever a fluid encounters an obstacle in its path, it has a problem to solve. Material that has arrived earlier must get out of the way of following particles, otherwise the fluid will build up in front indefinitely, eventually swamping the whole process. To avoid this, the fluid must contrive to flow around the obstacle and so escape away downstream. The action of flowing around involves motion perpendicular to the original flow, i.e. deflection. If the flow is not too fast, deflection can be accomplished in a gradual, orderly aerodynamic way where the presence of the impending obstacle is 'felt' upstream through the presence of waves—sound waves or Alfvén waves—acting as heralds of the encounter. Smooth, aerodynamic progress is then effected.

If the flow is too fast—supersonic and super-Alfvénic—the problem has to be solved differently. This is the situation prevailing in the solar wind. In this case material that has not actually collided with the obstacle, but which has undergone a preliminary transition in which it has become deflected, builds up upstream. The deflection takes place in a transition region known as a shock [40]. Such shocks are commonly bow-shaped in cross section and known as 'bow shocks'. Figure 2.3 illustrates the process. Passage through the shock causes the fluid to undergo an irreversible, entropy-increasing change which heats the fluid. In a fluid within which inter-particle collisions are important, these effect the necessary changes, but in a collisionless medium, such as the solar wind, the exchanges are effected via waves. The shock-heated solar wind forms a blanket or 'magnetosheath' of heated plasma around the obstacle. The sound speed increases from the solar wind to the magnetosheath because of the rise in temperature and the Alfvén speed increases because the magnetic field strength increases faster than the square root of the number density. The problem can thus be seen to have been overcome very neatly; the medium through which the solar wind is diverted has been modified to the degree necessary for waves to propagate between the obstacle and original flow. Shocks are naturally most pronounced directly upstream from obstacles, where the greatest degree of deflection is required. They curve around the obstacle and diminish in strength as the necessity for them reduces at greater distances from the sub-solar point or 'nose'.

The situation is not actually quite as clear cut as this, since some solar-wind particles will be reflected back upstream by the ramp in magnetic field strength from the solar wind to the magnetosheath, and some particles from the magnetosheath will find their way back out into the solar wind. The extent to which material and indeed some plasma waves find their way upstream and consequently reduce the severity of the shock depends on the direction of the solar wind's magnetic field. When and where the field is closely parallel to a line at right angles to the shock surface, the transition across this 'parallel shock' is relatively gradual. At a 'perpendicular shock' the transition is more

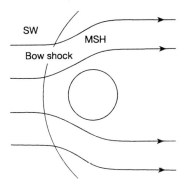

Figure 2.3. Deflection of the supersonic solar wind **SW** by an obstacle (open circle) involves the setting up of a bow shock and a sheath of modified solar wind around the obstacle. The sheath is usually referred to as a 'magnetosheath' **MSH**, whether or not the obstacle is a magnetic field. 'Ionosheath' is sometimes used, instead, when the obstacle is a planetary or cometary ionosphere.

sudden and more violent. Since the shock surface is curved, parallel shocks and perpendicular shocks can prevail at different places at the same time.

Since the electrical conductivity of the magnetosheath remains high, particles originally sharing a tube of force will continue to do so even though those passing close to the obstacle are slowed down. Continuity under these conditions causes magnetic lines of force of the sheath region to become draped around the obstacle, as indicated in figure 2.4.

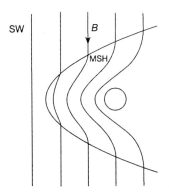

Figure 2.4. Draping of magnetic lines of force downstream from a bow shock. For simplicity, the figure is drawn for an interplanetary magnetic field normal to solar-wind flow. The magnetic field within the magnetosheath is seen to be draped around the obstacle.

The size of the enclave from which the heated solar wind, or magnetosheath, is excluded, known as a magnetosphere [41], depends on the strength and extent of the planet's intrinsic magnetic field. A measure of both the strength and extent of the magnetic field is the magnetic moment, a quantity which may be expressed in terms of magnetic field multiplied by distance cubed. Whatever the magnetic moment, of course, the field strength falls off quite rapidly with distance from a magnetised object. If the body is magnetised as a dipole, as with a bar magnet for example, the field strength weakens with the cube of the distance along any radial direction. The Earth's field, while it is more complicated than a dipole field close to the surface, approximates to that of a dipole at distances greater than a few Earth radii.

2.3.2   Unmagnetised bodies

Planets and other solar system bodies without magnetic fields also present obstacles to solar-wind flow [42]. In such cases the interaction is with the exosphere, ionosphere, atmosphere or the surface itself. The details of the interaction vary greatly, but many of the general effects just described still occur. In each case, the solar wind, being unable to build up indefinitely in front of the body, is forced to change in a way that will allow it to flow around. The sheath surrounding the unmagnetised obstacle is sometimes referred to as an ionosheath, but the term 'magnetosheath' is also quite commonly used for this region within which the magnetic field still departs markedly from its solar-wind condition, just as in figure 2.4. The change in properties between the solar wind and the magnetosheath may be more or less rapid depending on the circumstances. If the body is surrounded by an escaping gas, or exosphere, the solar wind may have considerable advance warning as it encounters atoms, electrons, ions, molecules and even particles of dust emanating from the body. A comet passing 'close' to the Sun, for example, releases copious material by sublimation of its ices as the result of solar heating and from solar ultra-violet radiation. The effects can be 'felt' by the solar wind as far as, in the case of Halley's comet, a million kilometres away [43]. The pick-up process mentioned above involves energy being transferred to the picked-up particles. Since the energisation is effected through the electric field, and the electric field, in turn, is derived from the solar wind, energy conservation requires the solar wind to lose as much energy as the picked-up particles gain. Loss of energy by the solar wind may in such cases take place gradually with the result that the eventual deflection around the object may also set in gradually. The shock in such cases may be relatively weak, or even absent.

2.3.3   Electrons of the Earth's magnetosheath

Examples of electron energy distributions taken in the magnetosheath, downstream from the terrestrial bow shock, are shown in figure 2.5. The magnetosheath distribution is often referred to as being flat at the lowest energies, or 'flat-topped'. The description seems to have arisen in an unusual way. The measurements leading to it [44] actually rose from a plateau towards low energies. However, the rise was thought to be due to a background of secondary electrons released from the spacecraft by solar ultra-violet radiation. It could not, therefore, be considered to be a feature of the magnetosheath electron distribution. Over the years, the plateau that was left in this and other experiments has come to be thought of as the high-energy portion of a 'flat top' extending down to zero energy. Subsequent measurements that we made with detectors less susceptible to background counts, by virtue of electrostatic energy filters with greater angles of deflection and therefore greater discrimination, have consistently observed a rising distribution which we believe to be real. The significance of the shape of the distribution in this region will become clear in Chapter 4 when we discuss the processes which may lead to its generation from the solar wind at the Earth's bow shock. The transition between the electrons of the magnetosheath and the electrons of the magnetosphere will be the subject of Chapter 7.

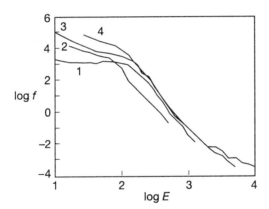

Figure 2.5. Electron energy distributions encountered in the terrestrial magnetosheath: **1** [45]; **2** [46]; **3** and **4** [47].

2.4 Magnetosphere formation

There are many formative elements and they are strongly interdependent. We shall attempt though, to introduce them and their main effects individually.

2.4.1 Magnetic field

We begin by considering a planet with an intrinsic magnetic field. Solar system planets with intrinsic magnetic fields are Mercury, Earth, Jupiter, Saturn, Uranus and Neptune. The fields of Venus and Mars are, if they exist at all, below the present limit of measurement, while the Pluto/Charon binary remains an unknown. In figure 2.6 the magnetic lines of force are those of a dipole. In the absence of other effects, which we will bring in gradually below, the lines of force all run between hemispheres—in one sense outside the planet, and in the opposite sense within. The extreme case of the line of force lying exactly along the axis will also do the latter, but via infinity. The pattern would extend out into space indefinitely. All dipole lines of force have the same shape, differing only in scale. The shape is given by

$$r = r_0 \cos^2 \Lambda$$

where r is the radial distance from the dipole centre, r_0 is the radial distance at the equatorial crossing point and Λ is the latitude. In order to draw lines of force at greater or lesser distances from the planet, it is necessary only to change the size of the diameter of the circle representing the planet.

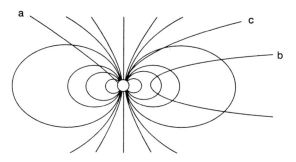

Figure 2.6. Deflection of cosmic rays by a planetary (in this case dipole) magnetic field. Cosmic rays with high enough energy or 'rigidity' **a** can penetrate the field at all latitudes. Lower-rigidity particles are deflected away at low latitudes **b**, but able to approach at high latitudes where they can travel relatively undeflected in the magnetic field direction **c**.

2.4.2   Cosmic rays [48, 49]

Interstellar space and possibly intergalactic space, too, is pervaded by cosmic rays. These protons, heavier nuclei and electrons which find their way to a planet's vicinity would, in the absence of the factors to be introduced, encounter the magnetic field just described. This field causes them to be deflected to a greater or lesser extent, depending on the position and direction of arrival and on the particle's species and energy. A convenient measure of a particle's ability to resist the bending of its trajectory by a magnetic field is its magnetic rigidity. Magnetic rigidity P is defined as momentum divided by the total charge (positive or negative), Ze, i.e.

$$P = \sqrt{E^2 + 2EE_0}/Ze \quad V$$

where energies are in eV. The gyro-radius r_c of a particle of rigidity P in a magnetic field whose component normal to the trajectory is B nT is,

$$r_c = 3.33P/B.$$

The magnetic field strength, or induction at radial distance r and latitude Λ is given approximately by

$$B = \mathcal{M}(3\sin^2 \Lambda + 1)/r^3 \text{ T}$$

where $\mathcal{M}$ is the magnetic moment, which for the Earth is 7.9×10^{15} T m^3.

In order for particles to be able to reach the planet's surface through its near-dipole magnetic field from any direction, a charged particle requires a rigidity of around 1.5×10^{10} V. Lower-rigidity particles, which may be deflected away at low latitudes, can approach from the polar regions where, although the field is double that at the equator at the same distance, the component normal to a near radial trajectory is much weaker and therefore unable to cause much deflection. The situation is summarised in figure 2.6.

When cosmic rays strike constituents of the atmosphere they produce ionisation and nuclear interactions. Some of the by-products are a splash, or albedo of charged and neutral particles returning upwards and outwards. By the same token that low-rigidity particles are unable to penetrate the magnetic field from outside, low-rigidity particles released inside cannot leave. These are guided by the field back to the atmosphere where as re-entrant albedo they add to the incoming flux of cosmic rays.

The neutral-particle albedo consists primarily of neutrons. Neutrons are unstable and decay with a half-life of 889 s into a proton, an electron and a neutrino [50]. This results in protons and electrons being deposited throughout the region, as illustrated in figure 2.7, with, naturally, higher numbers closer to the planet. Some of these offspring may find themselves guided straight back to the atmosphere by the magnetic field. Some may escape into space. Others, however, will be unable to do either of these and find themselves trapped in the field, as indicated here schematically, by a process to be looked into next.

2.4.3 Trapped particles [51]

Once deposited, a particle's motion in the magnetic field in which it finds itself consists of a gyration perpendicular to the field and a drift parallel to the field. If the field varies in strength along its length, the force exerted on the particle has, as shown in figure 2.8, a component directed away from the region of strong field. This serves to drive the gyrating particle in the direction of weaker magnetic field.

Another, and very convenient, way of looking at the effect is to note that the particle is repelled by the magnetic field gradient by virtue of the particle's magnetic moment. The force of repulsion is equal to the product of the field

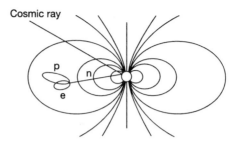

Figure 2.7. Cosmic rays, on interacting with the planet's atmosphere or surface, produce an outpouring albedo. The charged-particle component either escapes or is guided back into the atmosphere, as a re-entrant albedo, while neutrons **n** travel force-free until they decay into protons **p**, electrons **e** and neutrinos. The neutrinos escape into space while some of the protons and electrons are guided back to the planet and some become trapped by the magnetic field.

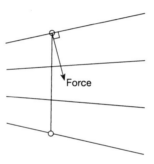

Figure 2.8. The Lorentz force on a charged particle gyrating (indicated in cross section by the open circles) in a magnetic field (converging lines) drives the particle from the stronger field towards the weaker field.

gradient and the magnetic moment, just as for any magnetic dipole such as a compass needle. The magnetic moment μ of a gyrating non-relativistic, charged particle is

$$\mu = E \sin^2 \alpha / B$$

or, including the effects of relativity,

$$\mu = p_\perp^2 / 2B\Gamma m$$

where m is the rest mass.

The force $\mathcal{F}$ is anti-parallel to the field gradient and is

$$\mathcal{F} = -\mu \, \mathrm{grad} \, B.$$

The consequent motion will in general be very complicated. However, it was demonstrated by Alfvén that under conditions where the particle executes many gyrations in moving between significantly changing magnetic field strengths, the particle's magnetic moment will remain constant. In view of this, the magnetic moment is known as the first adiabatic invariant of motion [52]—the term 'adiabatic' indicating that no energy is added to or taken from the system and all changes, temporal and spatial, are gradual relative to characteristic periods and gyro-radii.

Since kinetic energy is conserved in static fields we also have

$$\sin^2 \alpha / B = \text{constant.}$$

This shows that the pitch angle α, between the particle's trajectory and the local magnetic vector, increases with field strength. This it can do only until $\alpha = 90°$, at which point the particle is unable to penetrate further into the increasing field and is reflected back towards weaker field. A particle finding itself in a weak field between two regions of strong field will be forced to bounce back and forth if its pitch angle is, by these means, increased to 90° at any point or points, known as 'mirror points'. The particle is thus trapped in a magnetic 'bottle' as depicted in figure 2.9. If this condition fails to be met at either end, the particle penetrates through the strong field region into the atmosphere, or whatever lies below.

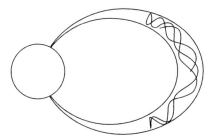

Figure 2.9. Gyration about, and motion along the magnetic field direction combine to produce a helix which gradually tightens as stronger fields are approached. The initial conditions determine whether a charged particle is repeatedly reflected to bounce between 'mirror points' (as here), or whether it penetrates to the planet's atmosphere or surface.

Since the magnetic field falls off with distance from the planet there is everywhere a gradient perpendicular to the magnetic vector. Figure 2.10 shows the type of trajectory followed by a charged particle, in this case an electron, gyrating in such a field. The motion, purely the response to magnetic deflection, is in the form of a drift of the point about which the particle gyrates—its 'guiding centre'. The drift is perpendicular to the field and to its gradient. The drift velocity is,

$$v_{d \text{ gradient}} = p_\perp^2 \operatorname{grad} B \sin \theta / 2Ze\Gamma m B^2$$

where θ is the angle between the field and its gradient. Protons and other charged particles drift in the opposite direction to electrons.

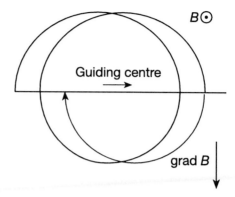

Figure 2.10. Motion of a positively charged particle in a magnetic field (directed into the paper) with a gradient, illustrating the 'gradient drift' of the guiding centre of gyration caused by the variation in gyro-radius. Kinetic energy is a constant of the motion.

Centrifugal force arising from curvature of magnetic field lines also introduces a drift. The magnitude of this is

$$v_{\text{d curvature}} = p_{\parallel}^2 / Ze\Gamma m B r_{\text{cv}}$$

where r_{cv} is the local radius of curvature. The drift is perpendicular to the plane containing the curved line of force.

Adiabatic motion consists, then, of

(i)   gyration about the local magnetic vector;
(ii) oscillation between mirror points in the northern and southern hemispheres; and
(iii) azimuthal drift caused by the gradient in field strength and curvature of field lines.

These three modes of motion are summarised in figure 2.11.

The Earth's radiation belts, and those of the other magnetised planets, consist of populations of particles for which these conditions hold well enough to keep them trapped for considerable periods.

Trapped particles behave as a diamagnetic material, by opposing the field that is deflecting them. This weakens the field closer to the planet and strengthens it at greater distances, thus stretching out the field lines to inflate the overall magnetic regime as depicted in figure 2.12. This distortion sets a limit on the number of particles that can be trapped. A balance is reached between storage of fresh particles and loss from a weakening field and other causes. A magnetised planet without an atmosphere, such as Mercury, still undergoes a similar loading of its magnetosphere through an albedo released directly from its surface.

A second contribution to field distortion arises from the fact that electrons and ions gradient drift in opposite directions—in the Earth's field electrons drift

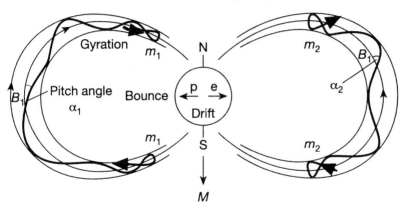

Figure 2.11. The three basic modes of trapped particle motion: gyration, drift and bounce. The figure applies to a planet which shares with the Earth a southward directed magnetic moment, M. Particles of pitch angle α_1 at a location where the magnetic field strength is B_1, mirror at points m_1 lying on a surface where $B = B_1/\sin^2 \alpha_1$. Particles with larger pitch angle α_2 at the same field strength mirror at higher altitudes, m_2. The locus of motion of each particle's guiding centre lies on a toroidal shell around the planet. Note that only those particles with energies low enough to meet criteria discussed in the text are able to execute this 'adiabatic motion' (from [57]).

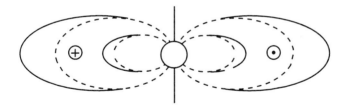

Figure 2.12. The trapped particles drift azimuthally to form a ring current around the planet. The ring current indicated in cross section is diamagnetic in the sense that its associated magnetic field opposes the original field inside the ring and enhances it outside. As a result the lines of force become extended outwards.

to the east and protons to the west—so there is a net current loop, the ring current, which also counters the parent field. Temporal changes in particle population and hence in the ring current give rise to readily measurable fluctuations in the geomagnetic field at the Earth's surface.

2.4.4 Planetary rotation

If the planet has an atmosphere, this will be caused by friction to co-rotate with the planet. Collisions between lower and higher atmospheric components will extend the friction to higher altitudes until they become too infrequent to do so at the steadily reducing number densities. A portion of the co-rotating atmosphere, the ionosphere, will contain electrons and ions resulting from solar ultra-violet and cosmic radiation. These particles will experience a deflecting force from the planet's magnetic field. In the situation depicted in figure 2.13, applicable to the Earth's equatorial plane viewed from the north, positively charged particles will be deflected outwards and negative ones inwards. The charge separation gives rise to an electric field countering the separation. The outcome is that charge continues to build up only until the force exerted by the electric field so created balances the deflecting force of the magnetic field. At this stage,

$$\mathcal{E}_q = -vB \sin \alpha$$

where α is the angle between v and B. Once this condition has been established, charged particles will experience no net force as long as they co-rotate with the planet. Freshly ionised particles of energy much less than that corresponding to the drift velocity will execute cycloidal motion whose mean azimuthal velocity is equal to the co-rotation velocity at the ionisation position. This behaviour parallels exactly that of particles being 'picked up' by the solar wind. The process gives rise to a region of the magnetosphere—actually an upward extension of the ionosphere—known as the plasmasphere.

Since the electric field $\mathcal{E}_q$ is at right angles to, not only the velocity, but also the magnetic field, it follows that co-rotation causes magnetic lines of force to become electrostatic equipotentials. This applies at all latitudes, so the electric field arising from co-rotation is directed as indicated in figure 2.14 on all field lines where co-rotation is in progress. In the case of a dipole where, at any latitude, magnetic field strength falls off with the cube of radial distance, while co-rotation velocity increases only linearly, the electric field falls off inversely with the square of radial distance at any latitude.

2.4.5 Magnetosheath plasma

We have, so far, built up a magnetosphere from the inside. What are the effects of the magnetosheath surrounding it and curtailing its outward extension? The primary effect is that the high electrical conductivity of the magnetosheath generates currents which exclude the planet's intrinsic magnetic field from the magnetosheath [53]. These currents at the same time modify the field within the magnetosphere. Lines of force are compressed on the dayside and stretched on the nightside of the planet, as sketched in figure 2.15(a). The magnetosphere takes the form of a comet-shaped object with a tail extending hundreds or even thousands of planetary radii downstream. Particles trapped within the magnetosphere are confined to regions where the magnetic field is 'well ordered' and changes relatively little with longitude or local time.

Although there are far too few collisions to promote a viscous interaction in the normal sense between the magnetosheath and particles trapped in the magnetosphere, these two media behave as though there were collisions, a

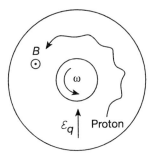

Figure 2.13. Co-rotation. Illustrated here in the equatorial plane of a magnetised planet with a frictionally dragged co-rotating neutral atmosphere is the process by which freshly injected charged particles or newly ionised particles are forced to co-rotate with the planet and its atmosphere. Any protons, for example, that happen to be co-rotating (e.g. as the result of ionisation of a co-rotating neutral) are driven outward, and negative charges are driven inwards by dynamo action until the resulting space-charge field $\mathcal{E}_q$ balances the induced field. Subsequently, charged particles injected into, or created within, the medium with zero velocity will initially be accelerated by $\mathcal{E}_q$—protons and other ions inward (as shown) and electrons outward—both species gaining kinetic energy in the process. Magnetic deflection will return them to the initial radial position and to their initial zero energy. The cycle will repeat, the net result being a cycloidal motion at the general co-rotation speed. Eventually, wave–particle interactions and/or friction will subsume the 'picked-up' ions and electrons into pure co-rotation.

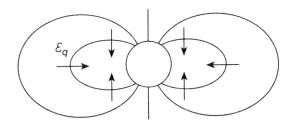

Figure 2.14. The co-rotation electric field $\mathcal{E}_q$ seen in a meridian plane. The field is everywhere perpendicular to the magnetic field, establishing magnetic lines of force as electric equipotentials.

phenomenon attributed to wave–particle interactions which effect the coupling. The trapped plasma is, as a consequence, set in sympathetic motion with the magnetosheath. The motion is naturally downstream close to the flanks, and there is a return flow towards the centre, in a two-cell convection [54, 55], as shown schematically in figure 2.15(*b*). The details of this flow are far from clear. We cannot even be certain whether the flow at any moment actually follows a closed circulation pattern which continuity requires only on average. The full

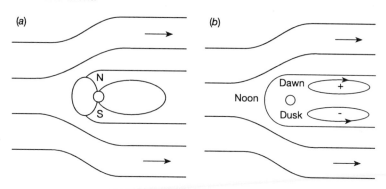

Figure 2.15. Containment of a planet's intrinsic magnetic field (inner lines) by the solar wind (direction arrowed). (*a*) Meridional view, illustrating the compression upstream and extension downstream. (*b*) View in equatorial plane, showing the downstream eddies formed by quasi-frictional drag of the solar wind, and the charge separation caused thereby.

three-dimensional picture can only be surmised at present, though circulation patterns in the ionosphere at the feet of the field lines threading the convection regions certainly indicate that the two-cell pattern prevails along magnetic field lines at all latitudes and altitudes down to the ionosphere.

This convection has a similar dynamo effect to that of the co-rotation just discussed, to solar-wind flow and to solar-wind pick-up examined earlier. Charged particles set in motion by the quasi-viscosity become deflected by the magnetic field, to set up, in the case of the Earth, a space charge, which is positive on the dawn side and negative on the dusk side of the magnetospheric tail. An electric field develops, as in the other examples, until it creates a restoring force to balance the deflecting force of the magnetic field. Once this stage has been reached, the particles can convect force-free. Potential differences between the 'eyes' of the dawn and dusk cells of the Earth's magnetosphere may amount to tens of kilovolts. Where convection maps down to the ionosphere, in the polar and near polar regions the potential difference is known as the 'polar-cap potential' or 'cross-pole potential'. The electric field arising from this motion is another key element of an environment being shaped by, as well as controlling, particle motion.

The actual motion of charged particles subject to the 'crossed' electric and magnetic fields is depicted in figure 2.16. The motion involves a cyclic variation in kinetic energy and a steady drift of the guiding centre. It is instructive to compare and contrast this '$\mathcal{E}$-cross-B' drift with the geometrically similar drift at constant energy caused by a magnetic field gradient.

If there is a magnetic-field gradient as well as an electric field the guiding centre drifts across equipotentials, adding to the cyclic variation in energy a systematic increase or decrease (depending on the sense of the field), as

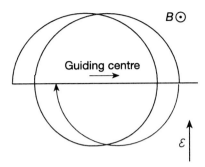

Figure 2.16. Motion of a positively charged particle in 'crossed' electric and magnetic fields, illustrating the '$\mathcal{E}$-cross-B' drift of the guiding centre resulting from a cyclic variation in kinetic energy and, consequently, in gyro-radius.

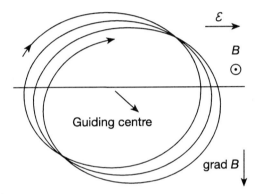

Figure 2.17. Motion of a positively charged particle in an electric field perpendicular to a magnetic field and to its perpendicular gradient. Superimposed on oscillatory changes there is a steady gain of energy as the guiding centre drifts into the region of stronger magnetic field with a component of velocity parallel to $\mathcal{E}$.

illustrated in figure 2.17.

2.4.6 Magnetosheath magnetic field

The magnetic field of the solar wind takes on a wide range of strengths and directions [56]. The directions 'expected' from considerations of the high electrical conductivity, its speed and the rotation of the Sun, lie in the ecliptic plane approximately 43° west of the Earth–Sun line, or at the complementary angle, depending on the magnetic polarity of the source region at the Sun. There are, though, wide fluctuations about this typical direction resulting from the interaction between different solar-wind streams, and waves and turbulence in interplanetary space. The field in the magnetosheath will, because of its link to the solar-wind field, be subject to these variations and will introduce further distortions of its own. The field that confronts the magnetospheric field at the boundary between the magnetosphere and the magnetosheath known as the 'magnetopause' can be expected therefore to take on, over a suitably long interval, the full range of directions. The juxtaposition of magnetosheath and magnetospheric fields will form a crucial element of Chapter 7. For the moment we just note that when or where the magnetosheath field aligns itself with the magnetospheric field the two regimes remain magnetically separate, as in figure 2.18(*a*) and that when or where the field directions are opposed magnetic nulls are created, which allow some lines of force to thread both plasmas as in figure 2.18(*b*). The fields are not required to be exactly opposed for this inter-connection to occur; it is necessary only that components of the fields are opposed. These latter configurations are believed to exist in limited regions and for limited times, depending on prevailing solar-wind conditions. At the dayside magnetopause these conditions are described as 'magnetic re-connection' when steady state, and as 'flux-transfer events' when temporary or sporadic.

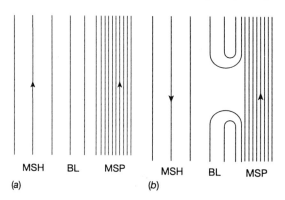

Figure 2.18.   Juxtaposition of magnetic fields of the magnetosheath MSH and magnetosphere MSP (*a*) when the fields are aligned, and (*b*) when the fields are opposed and create a magnetic null. BL is a boundary layer across which the transition occurs. More complicated transitions occur when the fields on the two sides of the boundary layer are skew, as they generally are.

Magnetic nulls almost certainly occur, too, in magnetospheric tails. In this case, the fields which cancel each other in a limited region of space are those which, as the result of the drawing out of the field, become opposed in the central plane of the magnetosphere. The type of configuration that is thought to be set up is shown in figure 2.19. There may be none, or one or more such magnetic nulls, depending on the state of the solar wind and its magnetic field. The dayside and nightside magnetic nulls may or may not be intimately connected.

Allowance for the magnetic field of the solar wind has, as we now see,

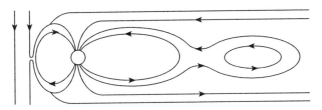

Figure 2.19. A schematic and simplified meridional cross section of a magnetosphere for a situation as in figure 2.18(*b*) where there is an upstream magnetic null. Adjacent opposing fields are thought to create a second null or magnetic 'neutral sheet' within the magnetosphere.

introduced into our model magnetosphere magnetic lines of force which, if they could be traced instantaneously would be found to permeate the magnetosphere, the magnetosheath and the solar wind. Lines intersecting the polar regions of the Earth appear also to belong to this category by escaping into the solar wind at great distances along the magnetospheric tail.

2.5   The Earth's magnetosphere

2.5.1   Configuration

Figure 2.20 shows the main features of the terrestrial magnetosphere as currently generally understood.   Far more is known, for obvious reasons, about the terrestrial magnetosphere than about any of the other planetary magnetospheres, all of which, with the exception of that of the Pluto/Charon binary system, have been visited by spacecraft. The terrestrial magnetosphere has been extensively investigated using rockets and satellites and by space probes on their way to extra-terrestrial destinations. High-altitude balloons, radio waves, photography and even unaided visual observations of aurorae have also played important roles in exploring its near-Earth regions.   It needs to be said, though, that despite the large number of missions that have taken place and despite the concentrated efforts that have been made to interpret the vast repository of measurements obtained, there are still major uncertainties about the configuration of the terrestrial magnetosphere and its workings [58].

The magnetosphere is dominated by the plasma sheet, a giant distorted torus. The torus is compressed on the sunward side of the Earth in sympathy with the compression of the magnetosphere on that side.   The torus is correspondingly extended downstream. Along the dawn and dusk flanks of the plasma sheet, and on its sunward surface, there is a boundary layer, a part of the magnetopause, which forms the transition between the plasma sheet and the primarily solar plasma of the magnetosheath. The plasma sheet is bounded north and south by a boundary layer—the plasma sheet boundary layer (PSBL)—which separates

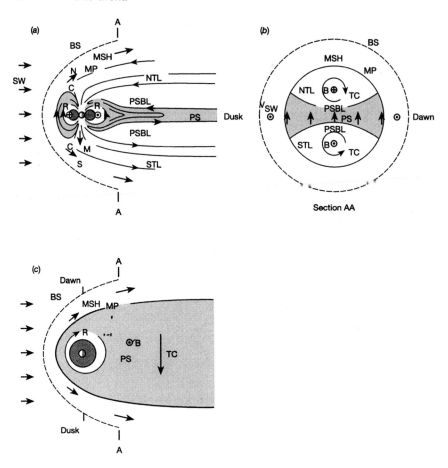

Figure 2.20. The magnetosphere of the Earth as currently envisaged: (*a*) noon–midnight plane perpendicular to the ecliptic; (*b*) dawn–dusk plane perpendicular to the ecliptic; (*c*) in the ecliptic plane. (*b*) is a cross section of (*a*) and (*b*) taken along **AA**. In order to avoid over- complication, the magnetosphere is drawn with the dipole tilted into or out of the paper. **BS** is the bow shock and **MSH** is the magnetosheath. **MP** is the magnetopause and **NTL** and **STL** are the northern and southern lobes of the magnetotail. **PS** (lightly shaded) is the plasma sheet and **PSBL** is its outer boundary layer. **R** represents the ring current and **TC** are the northern and southern elements of the tail-lobe current. The heavily shaded region is the plasmasphere. The cusps, **C**, are thought to permit ready access to and from the magnetosheath (from [57]).

it from northern and southern lobes of the magnetospheric tail, or magnetotail. It is unclear at present whether the PSBL is part of the same 'skin' as the magnetopause on the flanks, and the low-latitude boundary layer (LLBL) on the dayside [59]. The inner surface of the plasma sheet is the boundary between particles that co-rotate with the Earth as an outward extension of the low- and mid-latitude ionosphere, and particles of the plasma sheet which are governed by electric and magnetic fields of the magnetosphere.

Geomagnetic lines of force within the plasma sheet run connect points in the southern hemisphere to 'conjugate' points in the north. They can be thought of as the lines of force of a distorted dipole, compressed on the sunward side, or dayside, and extended away from the Sun on the nightside. These distortions are caused by electric currents in the magnetopause and the magnetosphere. Given any system of currents, it is a straightforward matter to evaluate the magnetic fields created by them: the reverse, going from effect to cause, is much harder, and it is the reverse we are faced with here. One thing is clear, though, and that is that the nightside extension of field lines within the plasma sheet, with oppositely directed field in the northern and southern halves of the plasma sheet, requires an electric current to flow around the magnetopause and through the central region of the plasma sheet. The central region of the plasma where the towards-the-Earth and away-from-the-Earth magnetic fields meet and cancel is known as the neutral sheet. The extent to which the neutral sheet and the current sheet coincide is still a matter of conjecture.

The plasma sheet is a constantly changing, dynamic entity. One of the reasons for this is that, closely linked as it is with the geomagnetic field, whose axis is inclined to the rotation axis, it naturally takes up the changing aspect of the field as the Earth undergoes its daily rotation. The innermost parts follow the changing tilt of the magnetic axis relative to the ecliptic, but the distant plasma sheet is like a pennant caught in a strong wind. Whether the far plasma sheet behaves as a single entity, as in the present-day picture shown here, or as a number of individual ragged filaments remains to be discovered. Another cause of constant change is the flow of particles within the plasma sheet. Plasma sheet dynamo motion serves to some extent to shape the magnetic field in the more remote (from the planet) regions. In this sense the behaviour is intermediate between that of the solar-wind particles which overwhelmingly control and shape the interplanetary and heliospheric field through which they flow, and the radiation belt particles, which by and large submit to the field in the near-Earth region which they occupy.

A prospective huge step forward in elucidating the whole picture met with a great setback when four identically instrumented spacecraft designed to operate in close co-ordination to give a much needed third dimension to magnetospheric studies came to grief in July 1996 with the catastrophic failure of Ariane V to launch the international Cluster mission on which many of the space plasma community's hopes were pinned. Fortunately, this mission has been resurrected and is to be launched at the turn of the millennium.

2.5.2 Temporal variations

The complexity that even the rudimentary considerations above promise for the magnetosphere is added to even further by that fact that this medium is almost continuously in a state of change. This is to be expected, if for no other reason than that the solar wind, a major controlling influence, itself constantly changes. Changes in solar-wind number density and velocity naturally vary the solar wind's ability to counter the geomagnetic field and thus vary the overall size of the magnetosphere. Changes in the solar-wind magnetic field—especially in direction—contribute very significantly, too, in changing the magnetic topology. Movements of boundaries, such as the magnetopause, propagate through the magnetosphere as magneto-hydrodynamic and other waves. These waves may excite other modes of disturbance, leading to a wide variety of oscillations able to interact with the constituent particles which, in turn, react back on and modify the waves.

The changes are not, however, all totally chaotic. One of the most striking facets of magnetospheric temporal change takes the recognisable form of a sequence known as a substorm [60]. The term 'substorm', although implying that it is in some way less than, or a mere component of, a greater entity, is really just a name for a sequence of events that may be recognised in the magnetosphere (a 'magnetospheric substorm'), in the aurora (an 'auroral substorm') or in any aspect of geophysics affected by the magnetosphere. Although the details of the changes occurring from one substorm to another vary greatly, it is possible in most cases to detect similar underlying trends, just as it is in musical compositions. Some substorms are in fact referred to as exhibiting 'classical' symptoms. For the magnetospheric substorm these are that, as the first stage, the plasma sheet which had become highly distended becomes thinner, at least in some regions. The next is that the overall magnetic field resorts to a more nearly dipolar structure. Whether this follows from the plasma sheet becoming so distended that it can no longer be contained and that the magnetosphere relaxes to a more stable configuration, or whether the sequence is responding directly to changes in the solar wind, or whether, again, the magnetosphere is responding to a combination of these influences is a matter of much current discussion. It is one of the main unresolved issues of magnetospheric physics today.

The auroral substorm [61], which is a direct result of changes occurring in the magnetosphere, also tends to follow a recognisable pattern. At the stage when the plasma sheet is distended, the aurora commonly takes the form of a quiescent arc [62] in the form of a curtain of luminescence stretching east–west, seen most clearly in the late evening sector at auroral latitudes. The beginning of a substorm is heralded by a sudden (within minutes) brightening of the arc or arcs. This brightening is followed often by striations of field-aligned rays [63] accompanied by movement of the arc and by its breaking up into separate rays, with other rays and patches [64] occurring over a wide (often whole-sky) area.

This stage of the substorm is well described as 'auroral break-up'. Break-up is seen most clearly from the midnight sector. Auroral arcs and break-up are both associated with electron acceleration, as we shall discuss in the next chapter. The last, and most nearly dipolar, stage of the magnetospheric substorm produces a patchy form of aurora with, very commonly, pulsations in brightness [63, 65, 66]. The electrons producing these patches appear to have been scattered from the plasma sheet and the radiation belts without acceleration, and the pulsations are due to variations in the rate at which these electrons are scattered from their trapped orbits into trajectories that precipitate them into the atmosphere [67, 68]. The cause of the pulsations is another major puzzle. The pulsating aurora is a phenomenon primarily of the early dayside sector.

2.5.3 Electron populations

2.5.3.1 Plasma sheet

The ions and electrons forming the plasma sheet have typical energies of 1 keV or slightly greater [69, 70]. Some example electron distributions are shown in figure 2.21. In keeping with the dynamic nature of the plasma sheet, its electrons are also highly variable. There is clearly no simple way to describe the distributions. They are certainly not Maxwellians, neither are they power laws. The more general kappa distribution meets with some success in some cases, but this cannot be claimed as a rule. The examples shown indicate (and this does seem to be a general rule) that the plasma sheet electrons have a greater proportion of higher-energy electrons, above 1 keV, than either the solar wind

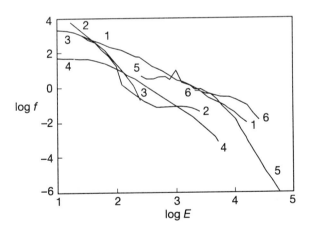

Figure 2.21. Electron energy distributions in the plasma sheet: **1** [47]; **2** [46]; **3** and **4** [71]; **5** [69]; **6** inferred from electrons producing a pulsating aurora [72]. Here and throughout the units of f are $km^{-6}\ s^3$. Here, and in all figures, the units of E are eV unless otherwise stated.

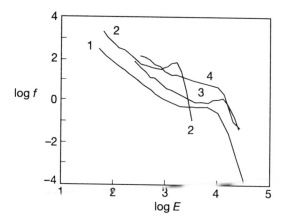

Figure 2.22. Electron energy distributions displaying features indicative of acceleration downward towards the atmosphere in the auroral zone: **1** [73]; **2** [74]; **3** [75]; **4** [76].

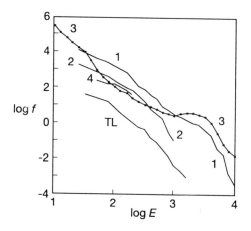

Figure 2.23. Electron energy distributions in the northern tail lobe **TL** [45], and precipitating over the northern polar cap: **1** and **2** [45], **3** and **4** [77].

or the magnetosheath.

Electrons precipitating from the plasma sheet and its boundary layers into the atmosphere where they cause ionisation and excitation are the main contributors to the terrestrial aurora. Some examples of auroral electron energy distributions are shown in figure 2.22. One of the puzzles we shall be trying to unfathom, in the next chapter, is what makes these electrons more energetic than those of the plasma sheet itself.

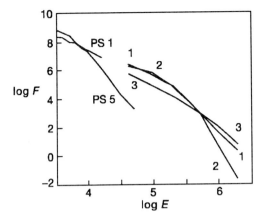

Figure 2.24. Electron energy distributions in **1** the inner radiation zone, **2** the 'slot' region and **3** the outer zone [79]. Plasma sheet distributions **1** and **5** from figure 2.21 are shown for comparison. In view of the relativistic nature of the radiation-zone electrons, the distributions are shown in terms of phase-space density F. measured here and throughout in units of $(zJ\ s)^{-3}$.

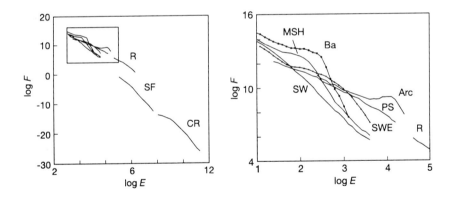

Figure 2.25. A panoramic view, with inset enlarged, extending over ten decades of energy and some 40 of phase-space density, of examples of electron energy distributions whose origin the following chapters will do their best to fathom. **CR**, cosmic rays [80, 81]; **R**, outer radiation belt [79]; **SF**, solar-flare electrons [82]; **Arc**, responsible for an auroral arc [75]; **Ba**, generated by barium ions released into the solar-wind [83]; **MSH**, in the magnetosheath [47]; **PS**, plasma sheet [47]; **SW**, solar wind [47]; and **SWE**, encountered during 'solar-wind events' [47].

2.5.3.2 *Polar cap precipitation*

The lone trace **TL** in figure 2.23 demonstrates the paucity of measurements in the tail lobes. The difficulty is that velocity-space densities are so low that they are below the resolution of detectors designed to look at other regions such as the plasma sheet, and as far as I know there have been no missions dedicated to this task. The particles here are even more diffuse than those of the plasma sheet. Magnetic lines of force within the tail lobes do not connect points in the northern and southern hemispheres (except within the Earth). Lines of force in these regions appear to connect with those of interplanetary space, though it is still unclear where they leave the extended tail of the magnetosphere to make this transition. Examples of electron precipitation from the tail lobes into the polar regions, where they produce the polar aurora, are also shown in figure 2.23. These are to be discussed in Chapter 8. For the moment we just note that the precipitation here is enhanced relative to the tail lobe, but generally rather weaker than that from the plasma sheet and its outer boundary layer.

Co-existing with the inner part of the plasma sheet are the zones of geomagnetically trapped, or Van Allen, radiation [78]. These are composed of charged particles, electrons, protons, helium, carbon, oxygen and other ions with a wide range of net positive charge, trapped by the geomagnetic field. Electron distributions typical of the region are shown in figure 2.24. The essential difference between radiation belt particles and their lower-energy fellow travellers of the plasma sheet lies in their different response, because of their different energies, to the electric and magnetic fields they encounter in their common region of space. The electric fields which control and are set up by convective motion are largely ignored by the higher-energy particles of the radiation belt.

2.6 Panorama

The survey is completed with a panoramic view, in figure 2.25, of examples from the full set of electron distributions whose origins the following pages will do their best to fathom.

Chapter 3

Electron acceleration in the aurora

The aurora makes an ideal starting point for our explorations. Acceleration is particularly well marked and well documented here and, occurring relatively nearby on just the threshold of outer space, there is a great wealth of information about the phenomenon from many sources. The issues and processes we shall discuss have a relevance to most of the other acceleration sites on our itinerary, so the discussion here will serve as a central reference for these, too. What makes auroral electron acceleration particularly intriguing is that it constitutes one of the major puzzles confronting space plasma physics today.

We begin with a brief description of the aurora, recognising that using words and pictures alone makes this a near-impossible task for so complex and varied a phenomenon. We shall then look at the observations—the facts of the case—to see what we are dealing with and what the immediate inferences are. The 'established' potential-difference theory of electron acceleration is then described and discussed. This is followed by a parallel treatment of a new, wave theory which my colleagues and I developed upon becoming convinced that the established theory could not possibly be correct.

3.1 The aurora

The aurora [84, 85] is one of the most spectacular wonders of the natural world. It consists of lights in the sky that are most commonly seen in bands of latitude some 20° from the geomagnetic poles. On rare occasions, aurorae have been observed at latitudes right down to the equator [84]. The northern and southern counterparts, the aurora borealis and aurora australis, have been shown sometimes to be almost mirror images of one another [85]. Aurorae occur on other planets, too, including Jupiter, Saturn and Uranus [86]. The aurora can be perceived and appreciated from many different standpoints.

3.1.1 As a spectacle

The aurora takes many different forms. At its most bland it can exist as a more-or-less even and steady glow stretching to the horizon in all directions. The intensity of the light in this form of aurora is usually rather low, being close to the limit of detection by the unaided eye. It is only really detectable on a dark clear night, and even then is easy to miss. A tip to help distinguish a glow aurora from wispy moonlit cloud is that while clouds tend to obscure the stars, the aurora does not. Photographs of this form, although it is a common one, rarely, if ever, make their way into books since all we see are the stars against a background that is not quite black. In fact the colour is usually a green which, because it is faint and colour perception is poor at low intensities, seems to be a light grey. A notable exception occurs when the glow is a quite startling red, as it was on the famous occasion when the Roman army was despatched to deal with what was thought to be a great fire due to a battle in a city to the north [84, 86].

The glow aurora is sometimes quite patchy, and the patches more often than not fluctuate in brightness. The patches are irregular and fill a field of view of 5° or so, about ten times the angular size of the Sun or moon. When studied carefully they may be seen to pulsate often nearly regularly with a typical period of some 10 s. This, again, is easy to miss unless you gaze patiently at one spot for the required time. Nearby patches seem to pulsate sympathetically, giving the whole tableau the appearance of a collection of hot coals brought repeatedly into life by a gusty breeze. Figure 3.1 shows a typical pulsation cycle. Glows and pulsating patches are seen most commonly in the early morning hours.

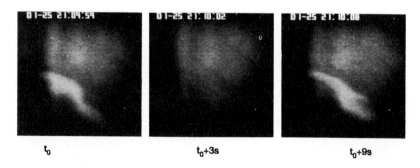

t_0 t_0+3s t_0+9s

Figure 3.1. A pulsating aurora recorded on 25 January 1979 by an image-intensifier television camera pointing along the magnetic field direction at Kiruna in northern Sweden. The period of this pulsation is 9 s. The sides of the frame correspond to horizontal distances of 90 km at the auroral altitude of 110 km [66].

The most spectacular form of aurora, the one we shall be particularly interested in in our study, is the auroral arc [62]. While this is a splendid name for a most arresting phenomenon, the name is actually rather misleading.

What is meant by the term is the luminous curtain stretching, over possibly more than 1000 km, usually between the eastern and western horizons. Perspective causes the curtain when quiescent to look something like a great arch, from which the term derives. Figure 3.2 shows a typical quiescent auroral arc over the Norwegian sea against the setting Sun. True to its 'curtain-like' description an arc is very much thinner in its north–south extent, at times appearing to be just 1 km thick. An arc may consist of several, roughly parallel, apparently separate thin curtains, as here, and may contain many giant folds, as in the extracts from an all-sky photograph of a multiple arc on the front cover. The light emanating from convoluted folds in auroral arc curtains sometimes creates the impression of vertical, or near-vertical, stripes. Sometimes the curtain is genuinely, intrinsically, striped. Detailed analysis of photographs shows these stripes or 'rays' to be aligned with the magnetic field. Auroral arcs can be bright enough to use as a reading lamp, although it would be something of a waste to use it as such, since the aurora is vastly more interesting than any document (even this one). There may be a glow aurora taking place at the same time on the equatorward side of an arc. The sky on the poleward side is usually very much darker. This is very apparent in the cover's central panel (from [89]).

Figure 3.2. An auroral arc photographed from Andøya in northern Norway with the setting Sun on the horizon. The arc appears to be composed of a number of distinct curtains of luminosity in a generally east–west direction. Rocket and satellite flights through the streams of electrons responsible for such forms show the precipitation to be continuous, with variations in energy and intensity across the whole band.
[Photograph T Edwards; see also [89].]

While an auroral arc might last for tens of minutes, possibly an hour or more [87], quite frequently it drifts gently towards lower latitudes during this time. Part of the motion may be real and part may simply be due to the rotation of the Earth on its geographic axis under a pattern fixed in space relative to the tilted magnetic axis. The arc may just fade away after this, or it may develop into perhaps the most spectacular of nature's sights, the 'auroral break-up'. In such an event the arc will quite suddenly become even brighter and more convoluted. The folds may break away as separate rays, and new rays may appear briefly in a riot of changing patterns and colours over the whole sky. An auroral break-up is quite breathtaking and gives rise to the very real possibility, if it takes place directly overhead, of the observer getting a twisted neck trying to follow what is going on. The best if not the only way of capturing the bewildering sequence of events is to just lie on the ground, assuming you are well wrapped up against the arctic or antarctic cold, gazing upward for the minutes this phase may last, trying to take it all in. The break-up will be another of our key phenomena. Still photographs, unfortunately, give no real impression of this dynamic type of aurora, but some of the films and videos (e.g. [88]) that are now available serve well to whet the appetite for a first-hand unforgettable experience.

The sequence of events, beginning with a quiet arc, often in the early evening, and continuing into break-up at around midnight, commonly ends with the sky being filled with pulsating patches well into the early hours of the morning. Such a sequence is described, as mentioned in Chapter 2, as an auroral substorm. The latter term, despite its prefix, is, as we saw, best seen as a term in its own right, and not as a minor or partial version of a greater entity.

It must be emphasised that the above provides only a thumbnail sketch of an exceedingly complex phenomenon. It will, though, give us something of a visual framework for the investigations to follow.

3.1.2   As a geophysical phenomenon

We have said nothing so far about what is actually going on to produce the aurora and what significance the aurora might have for the planet as a whole. As intimated in Chapter 2, the aurora occurs at the interface between the atmosphere, composed predominantly of neutral particles, and the magnetosphere, which is a plasma of charged ions and electrons, not to forget the major component of neutrals. The basic features of this aspect of the phenomenon are well understood. Charged particles, primarily electrons, with kinetic energies of several keV, stream from the magnetosphere into the atmosphere through the intermediate region of the ionised upper atmosphere or ionosphere. As they encounter rapidly increasing particle number densities they collide in greater and greater numbers with the constituent atoms and molecules, losing kinetic energy in the process. The atoms and molecules of the atmosphere are affected not only by the transfer of kinetic energy. They may also find their electrons raised to higher energy levels thus raising the atom or molecule into an excited state from which it can recover by emitting light of wavelengths characteristic of the atom, ion or molecule. Some collisions will strip electrons from the original neutral constituent thus leaving it as an ion or an ionised molecule. The ion or molecule may emit radiation characteristic of its current state, or emission may be delayed until the particle collects an unsuspecting electron, to become neutral again. The wide diversity of particles forming the atmosphere, the variation in composition with altitude and the almost infinite number of permutations of collisions transformations and recoveries that can occur combine to give the aurora a very rich spectrum of colours. The dominant emissions from electrons of several keV able to penetrate down to 150–100 km before running out of energy are a green line at 557.7 nm from excited oxygen and a red line at 661.1 nm from nitrogen. Electrons of 1 keV are stopped at altitudes $\sim$ 200 km where oxygen atoms predominate, to produce a red line of 630 nm. A red line of 656.3 nm can also be emitted by precipitating protons that have acquired an electron to become excited hydrogen atoms. Perhaps it was one of these events that the Roman generals saw.

The net result of all this is that the precipitating auroral particles generate a very significant quantity of ionisation in the slowing-down region of, say, 200 km to 100 km, known as the lower F-region and E-region of the ionosphere. Electron number densities thus caused may be as high in the nighttime ionosphere as those produced during the day by photo-ionisation from solar ultra-violet light. The rate of input of energy may be more than 200 mW m^{-2} locally and 10^{10} W overall, the latter being the equivalent of ten large power stations and a good fraction of the average electricity demand of the whole of the United Kingdom. This is an important source of heat for the upper atmosphere, certainly at high latitudes and possibly eventually for the atmosphere as a whole. The auroral input must therefore be taken into account in models of global energy balance.

3.1.3 As a plasma physics laboratory

The aurora affords mankind a splendidly diverse and dynamic laboratory for plasma physics. The processes taking place here naturally accomplish many of the feats we should like to emulate on the ground, and they do so on such a grand scale that we can investigate the details with a resolution that is quite impossible in the confined spaces of engineered plasmas. If you think of our spacecraft as Gulliver [90] exploring Brobdingnag, the land of the giants, you will be on the right track. We saw earlier that in many ways the plasmas of space may be seen as giant versions of the smaller-scale plasmas employed in the laboratory. These natural plasma phenomena have, therefore much to teach us about the key issues of plasma confinement, plasma heating, plasma instabilities and of course particle acceleration. We will take up this aspect now with special reference to our brief to find what it is that causes electrons to accelerate as they stream from the magnetosphere into the atmosphere in the Earth's auroral zones.

3.2   Observations of auroral electrons

The facts are not at all in dispute. An extensive collection of measurements has been acquired painstakingly and with steadily improving resolution over the past 35 years or so by many research teams using rockets and satellites to traverse the auroral zone. Stated briefly they are that the electrons responsible for some types of aurora, generally speaking the brighter and more highly structured, or discrete, types, such as auroral arcs, are accelerated into the upper atmosphere during the final 10 000 km or so of their plunge from the magnetosphere. Furthermore, the acceleration of these particles is directed principally, possibly exclusively, downward along the magnetic field direction (parallel to the magnetic field in the northern hemisphere, anti-parallel in the south).

3.2.1   A demonstration

To my mind the clearest demonstration that electrons are accelerated into the auroral atmosphere comes from an ambitious and highly successful experiment involving two spacecraft. The spacecraft were the Dynamics Explorers 1 and 2. DE-1 crossed the auroral zone at high altitude (often at more than 10 000 km) while DE-2 patrolled below at around 700 km. The hope was (for it could not be guaranteed) that there would be occasions when the two spacecraft simultaneously crossed regions connected by the same magnetic line of force, i.e. regions that were magnetically conjugate, at a time when the lower spacecraft was detecting electrons typical of those responsible for discrete aurorae. Exact conjunctions could not be expected. It could reasonably be hoped, though, that there would be at least a few occasions when the aurora was widespread enough and stable enough for the assumption to be made that the same electron stream was sampled at two stages of its development. Fortunately, for this purpose the auroral arc, one of the most interesting and characteristic forms of discrete aurora, provided, if not an easy target, at least a practical one. The uniformity and stability could be checked from cameras looking down from the spacecraft. Several rendezvous of this kind were made and some priceless measurements were obtained.

Figure 3.3 compares energy distributions of electrons travelling downwards approximately parallel to the magnetic field, as observed at altitudes of 11 900 km by DE-1, and at 769 km by DE-2, magnetically conjugate to the same auroral arc [74]. This conjunction—or, more strictly, acceptable near-conjunction—occurred on 4 November 1981 in the evening sector of the auroral zone. The satellites crossed parts of the electron stream producing the same east–west band of aurora just over 2 min apart and just over 300 km distant horizontally, in an east–west direction. The images obtained from the cameras confirmed that no major changes occurred during this time or over this distance.

The electron energy distribution changes greatly between the two altitudes. The peak seen at the lower altitude is highly characteristic of the discrete aurora,

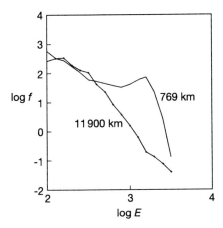

Figure 3.3. Clear evidence that electrons precipitating from the magnetosphere into the auroral-zone atmosphere are accelerated over the final 10 000 km or so of their plunge. The figure shows electron energy distributions from the same stream measured simultaneously at two altitudes by the satellites DE-1 and 2. The originally monotonic distribution has evolved into one containing a pronounced a peak at about 2 keV (derived from [74]).

though the energy where the peak occurs (1.6 keV) is relatively low. In other instances to be examined the peak energy sometimes exceeds 10 keV. The record, as far as I am aware, is for the energy flux, or count rate, to peak just above 30 keV [91].

The first question we need to ask is, can we take these measurements at face value? Can it be demonstrated, for example, that under the same conditions the two instruments would record the same thing? Some reassurance can be given on this since, when the spacecraft were earlier both above a diffuse and widespread glow aurora, where all indications are that there is little if any significant acceleration, the two instruments returned exactly the same spectra, at least at energies above 70 eV. Below 70 eV the DE-1 detector counted at a much faster rate that the DE-2 detector, both in the control and in the experiment itself. It seems prudent, therefore, to limit comparison to measurements above say 100 eV, where the present plot has its lower limit.

Except at the lowest energies, the DE-2 curve lies above the DE-1 curve. It seems, therefore, that somewhere between the two altitudes energy has been added. Even bearing in mind the uncertainties inherent in such a difficult experiment, which do not allow the results to be taken too literally, it is evident immediately that the acceleration process is not simple. Neither a uniform increase of energy, nor a uniform factor of increase, could account for the transformation.

Another, similar, event yielded at the altitude of 680 km a peak at 4 keV which had not been evident at 13 000 km. The departure from exact coincidence this time was 4 min and 460 km, but still confirmed by photographic evidence as being acceptable. A third conjunction, this time a more complicated one, showed acceleration taking place between altitudes of 13 000 and 700 km. A fourth event with a miss this time of just over 9 min and just over 400 km showed at around 14 000 km a pronounced peak at an energy of 3 keV indicating that, since such peaks are not seen in the magnetosphere, and irrespective of the accuracy of any conjunction, acceleration had occurred above this altitude. At 590 km altitude the peak was just 2 keV higher, suggesting, on the face of it, that acceleration had been, on that occasion, at least as strong—nominally stronger—above the altitude of 13 000 km than below.

What we have, then, is a series of uniquivocal demonstrations of acceleration occurring sometimes, but not always, below an altitude of 13 000 km. While this is enough to establish that there is a problem to solve, it is not enough to go on, so we seek more information. We shall begin by finding out more about the 'low-altitude' electrons, and then work our way up to the high-altitude distributions.

3.2.2   At low altitude

By 'low altitude', we mean here altitudes generally below the acceleration region, i.e. below 1000 km.

3.2.2.1   Characteristic peak

Figure 3.4 shows two energy distributions taken during the progress of a Skylark rocket launched in a northerly direction from Andøya in northern Norway across an auroral arc on the evening of 30 October 1973 [75]. One of the distributions was encountered at the southern edge of the arc, soon after measurements began. The peak in the distribution was at 4.5 keV. The other was taken 90 s later over the centre of the arc. The peak was then at the highest energy encountered during the flight, 13 keV. It is very noticeable here that despite the great differences at the higher energies, velocity-space densities at low energies are more or less the same at both locations. This invariance of the distribution at energies well below the main 'action', wherever it occurs, is a fairly consistent feature of auroral electron streams [92].

The extensive high-energy tail of the centre-of-arc distribution demonstrates that the processes at work can influence electrons up to energies of at least 30 keV. Other studies show that the effect extends up to at least several hundred keV [91], i.e. into the relativistic range.

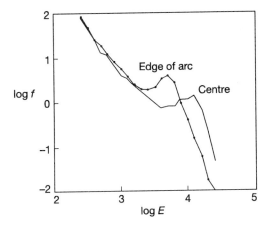

Figure 3.4. Electron energy distributions at the centre and edge of a stream producing an auroral arc. Both distributions exhibit peaks in velocity-space density, the peak at the centre occurring at higher energy but lower velocity-space density. The range of pitch angles sampled was 25–35° (from [75] and based on a Skylark rocket flight made from Andøya on 30 October 1973).

3.2.2.2 *Distributions without a peak*

A peaked distribution is not *de rigeur*, even for the brightest of discrete aurorae. We discovered this for ourselves when we were fortunate enough to launch a rocket to bring it conjugate with an auroral arc just as the arc broke up into bright rays rippling along its length [76]. The rocket was a Nike Tomahawk launched from Andøya in Norway near midnight on 27 November 1976. Between 280 and 290 s into the flight, with the rocket at an altitude of 200 km, it was engulfed by the stream of electrons producing lower down some bright rays. The distributions of figure 3.5 represent electrons of zero pitch angle within the rays and in the gaps between the rays but still within the arc. Both distributions are the averages of several ins and outs during the 10 s interval. Velocity-space densities fall off monotonically in both cases towards higher energies even though they were the causes of the brightest aurora seen on the flight and are in the process of carrying into the atmosphere far more energy (almost 200 mW m^{-2}) than unaccelerated plasma sheet electrons possibly could. This bulge or sloping shelf rather than a peak in what is clearly a strongly accelerated distribution will need to be accounted for.

3.2.2.3 *Inferences so far*

What general impressions and inferences can we gain and draw from the evidence so far? In all but the DE measurements, we are hampered by being unsure whether in the comparisons made between different parts of the same

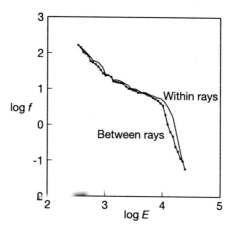

Figure 3.5. Electron energy distributions measured within and between rays in an auroral arc, observed from a Nike Tomahawk rocket launched from Andøya at 2051 UT on 27 November 1976. This demonstrates that the rays were produced by enhancement of acceleration rather than increases of flux [76].

stream we are viewing effects of the same source accelerated by different amounts, or whether we are witnessing the same accelerator acting on different sources, or even whether we are seeing a measure of both. While the last is certainly the safest assumption, it would perhaps be over-cautious to adopt it from the beginning. It would, in any case, leave us in the uncomfortable position of trying to explain why, if the sources for the two spectra in each pair were different, the velocity-space densities were so similar at low energies on all occasions examined so far. We shall adopt the working assumption that the source distributions are the same for both members of each pair. What we can never know, unfortunately, is whether the accelerator produces the weaker of the pair at an earlier stage of evolution, or whether the different degrees of acceleration occur on different tracks, so to speak. Even with this uncertainty, though, there are some important conclusions to draw. The first is that since the low-energy electrons are the same in each pair, despite the fact that the higher energies have changed greatly, the low energies are unlikely to be a secondary result, or by-product, of the higher energy ones. A second inference is that the accelerator, whatever it is, affects electrons of different energies differently, implying that acceleration involves a measure of resonance.

3.2.2.4 Field alignment

Figure 3.6 compares the velocity-space density at pitch angles of 0–10° with that at 50–60° during a rocket flight through an electron stream producing what was described as 'an active aurora' [93]. While the characteristic peak appears at both angles, it is considerably more marked at the smaller pitch angles where it also attains a higher maximum velocity-space density. There is not always, though, such a strong field alignment. In fact, the cores of stable structures rarely exhibit this feature. In the example of figure 3.7 velocity-space densities at the peak are the same for electrons travelling almost perpendicular to the magnetic field as they are for those moving nearly parallel. Even here though, the peak is slightly more pronounced at the smaller angle on account of the trough on its low-energy flank being deeper.

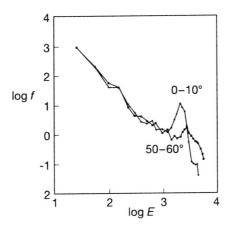

Figure 3.6. An auroral electron distribution exhibiting a higher and sharper peak at small pitch angles, 0–10°, than at medium pitch angles, 50–60° (adapted from [93]).

3.2.2.5 *Further inferences*

All indications are, then, that the accelerator works preferentially parallel to the magnetic field. The only way to avoid this conclusion would be to keep open the option that the initial, source distribution was also peaked in the field direction. This, however, would only be to put off the problem of explaining how that came about. Since the increasing field strength encountered during precipitation automatically increases pitch angles, any assumed primordial field alignment would need to be very much stronger than that observed [94]. The working assumption must be that we are witnessing field-aligned acceleration. The accelerator might be, not just preferentially, but wholly field aligned.

3.2.2.6 *Multiple peaks*

A feature which puts great demands on any interpretation is a double peak, e.g. [95]. The example shown in figure 3.8 [93] very typically has the peak at the higher energy approximately isotropic, while the peak at the lower energy

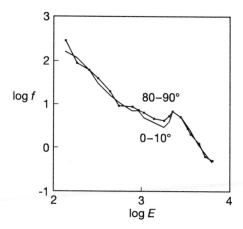

Figure 3.7. An auroral electron distribution with the same peak velocity-space density at small, 0–10°, and large, 80–90°, pitch angles, but with the peak at the smaller angles slightly more pronounced (adapted from [93]).

is strongly field aligned. It is conceivable that the lower peak arises from a different type of process to that producing the peak at higher energy, but the almost geometric similarity apparent when plotted on a log–log scale, as here, gives no real cause for believing that the two peaks are actually different in kind. The double peak provides a very tough test indeed for any would-be theory.

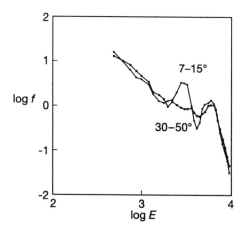

Figure 3.8. A double-peaked distribution, with a near-isotropic peak at high energy and a field-aligned peak at low energy (adapted from [93]).

3.2.2.7   Upward-flowing electrons

What happens to electrons flowing upwards through the acceleration region, away from the atmosphere? For example, are they retarded when the downward-flowing electrons are accelerated? Direct comparison is made complicated by the fact that the sources, and consequently source distributions, of the two families are different. The source of downward electrons is one of the magnetospheric plasmas such as the plasma sheet, the tail lobe, or the plasma sheet boundary layer between them, while the source for upward electrons is the ionised upper atmosphere or ionosphere. Electrons precipitating into the atmosphere suffer not only a loss of energy but are also strongly scattered in direction. Electrons resulting from the ionisation of atmospheric constituents are also a source of upflowing electrons.

Figure 3.9 shows an important and often quoted comparison of upgoing and downgoing electrons. The downgoing electrons exhibit here what might be termed a plateau, i.e. a form intermediate between the peaks and the bulges or falling shelves seen earlier. The plateau extends from approximately 1 keV to 3 keV. The upgoing electrons do not show a peak. At energies above about 500 eV velocity-space densities are considerably less than for the downgoing ones. At lower energies, though, velocity-space densities are approximately the same for both.

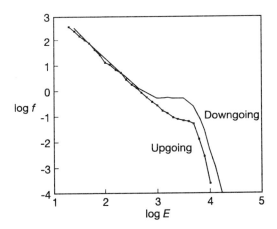

Figure 3.9. Comparison of precipitating (downgoing) and returning (upgoing) auroral electrons. Downgoing velocity-space densities exceed the upward ones at energies above 1000 eV, but at low energies the flow is approximately equal in the two directions (from [73]).

3.2.2.8 *Velocity-space distributions*

A broader picture can be gained by looking at the full velocity-space distributions that can now be acquired with more modern instruments and data gathering and processing techniques. Figure 3.10 shows at the lowest velocities a fairly isotropic distribution, slightly favouring downward directions, but at higher velocities there is a marked field-aligned bulge. At pitch angles above approximately 45° there is a broad plateau which becomes progressively less extensive for the upgoing electrons. At the altitude of the measurement, electrons with pitch angles of 71° or less are destined to plough into the atmosphere. Those with pitch angles between 71 and 90° will be mirrored by the magnetic field to occupy the range 109–90°; while those between 109 and 180° have emanated from the atmosphere and are just setting out on an upward journey. There is a sharp fall in velocity-space density at the end of the bulge, where the energy is approximately 10 keV, and the limit of measurement is reached at just over 15 keV in the parallel and perpendicular directions. Distributions with a similar plateau but without the field-aligned bulge have also been reported [99].

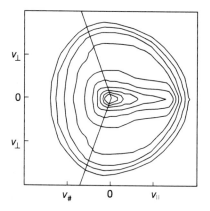

Figure 3.10. An electron distribution in an auroral arc elongated in the magnetic field direction. Velocities in the parallel (‖) anti-parallel (#) and perpendicular (⊥) directions range up to 7.5×10^7 m s^{-1} (corresponding to approximately 17 keV). Contours of velocity-space density are spaced approximately logarithmically at two per order of magnitude from 10^3 at the lowest velocities (centre) to 3×10^{-2} (outermost). The (upward) loss cone is indicated by the straight lines. The measurements were obtained from a Terrier Malemute rocket flight made from Poker Flat, Fort Yukon, Alaska on 3 February 1984. From [96]; see also [97] and [98] for many similar examples.

The distribution of figure 3.11 observed just 1 min later than that of figure 3.10 shows a crescent-shaped field-aligned peak. This is at an energy of 9 keV. Elsewhere, there is a broad plateau out to an energy of 10 keV in all forward directions, but more limited for (upgoing) electrons within the loss cone 68°

wide at this slighter higher altitude. Distributions similar to this but with the crescent extending along a circle of constant energy to complete a circular ridge or rampart from pitch angles of zero to beyond 90° have also been observed [99].

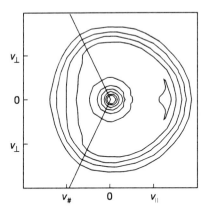

Figure 3.11. Electron distribution at the edge of the same auroral arc as in figure 3.10 showing a peak in the magnetic field direction. Other details as for figure 3.10 (from [96]).

3.2.2.9 Latitudinal structure

To complete the summary of the properties of auroral electrons at low altitudes we shall examine latitudinal cross sections of streams of electrons producing auroral arcs. Plate 2(*a*) is an energy–time spectrogram of electron velocity-space densities encountered on a flight through an auroral arc stream by a Skylark rocket [100]. Measurements extended over a distance of approximately 600 km in a northward direction, while the rocket altitude rose from around 200 km to just above 700 km, and back again.

For the first 120 of the 500 s of measurement the electrons are reminiscent of those found in the plasma sheet. There is no indication of acceleration in this period.

After a spell of very low velocity-space densities there is a stream of electrons accelerated to a degree producing a broad peak initially at around 2 keV then rising to 8 keV. The peak falls back to 2 keV and then rises again to remain at around 8 keV before disappearing quite suddenly. The second half of the sequence shows relatively low velocity-space densities, with one or two bursts confined to low energies.

An expanded cross section of the stream of accelerated electrons is shown in plate 2(*b*), where the colour-coded quantity is, this time, intensity. This variable exhibits a much more evident peak, and its movement over the energy range can be followed more readily.

The above sequence of events is typical of that experienced on long-range rocket flights or satellite passes of the auroral zone [101]. Sometimes, the central section with a distinctive peak in intensity or count rate is more orderly than that appearing here, the rise and fall of the energy at which the peak in intensity, count rate or (more rarely) velocity-space density occur having the form of an inverted letter V. For this reason the spatial structure has become known as an 'inverted V' [101].

The interpretation of plate 2(*a*) and its like is not absolutely straightforward. This is primarily because of the spatial-temporal ambiguity, whereby an uncertainty arises over whether changes observed from a moving vehicle correspond to traversal of a static spatial pattern, whether they are temporal changes occurring simultaneously, over the whole region or whether they are a combination of both. In this instance, we can be certain, as witnesses and from photographs, that at least a major part of the variation was caused by the crossing of a spatial structure. Some temporal fluctuations may also have been superimposed. On other occasions, an instrumented package has been ejected from the main payload to help resolve the ambiguity between time and space, e.g. [102].

Some of the features already noted may be seen here in this wider context. The relatively slight change at low energies in response to the varying position of the peak is particularly clear, as is the greater extension of the high-energy tail as the position of the peak rises. Plate 2(*a*) is fully consistent with the often found fact, seen in the earlier whole-sky photograph, that the region of acceleration is located between unaccelerated electrons on the equatorward side, and distinctly weaker precipitation on the poleward side.

One of the many surprises encountered by those trying to understand the aurora has been that, although the discrete aurora quite commonly appears to the eye as being composed of a set of quite narrow ($<$ 1 km wide) forms, the region where accelerated streams are found at rocket and satellite altitudes is typically some 50–100 km across. The apparent discrepancy has taken on the proportions of quite a dilemma. It will have been noticed that the so-called dilemma lacks a quantitative comparison. Let us attempt one now.

If we make the working assumption that the observed variations were due to a spatial structure within the arc stream, we can apply a test to see whether under the best possible viewing conditions, i.e. looking directly up at the arc along the local magnetic field direction, the arc would appear as a single or multiple structure. The scheme we adopted was to plot as a grey scale the energy flux carried by the electrons and thus a measure of the brightness of the aurora they produced. If the eye judges the relatively small variations (less than a factor of three) to be significant, the arc would be perceived as a multiple one. This certainly seems to be the case here, indicating that the elements of what appear to be multiple arcs are not necessarily produced by separate acceleration regions. Instead, they may be due to variations within a single overall region

[103].

Two further samples of electrons streams producing auroral arcs appear in plates 3(*a*) and (*b*). The detector from which plate 3(*a*) was drawn was set at a relatively large angle to the rocket in order to scan through a wide range of pitch angles during each 1.4 s spin of the rocket. The vertical stripes are produced by the repeated sampling of high intensities at small pitch angles, revealing very clearly the field-aligned nature of the precipitation. Plate 3(*b*) gives an indication of the great complexity that can be encountered (this is the occasion on which an arc developed the rays whose energy distributions were seen earlier).

The character of electrons below the acceleration region is vividly illustrated by the 'mountain' plot of figure 3.12 [104]. This represents a velocity distribution measured in an auroral arc and exhibits the characteristic field-aligned peak, the feature presenting the foremost challenge.

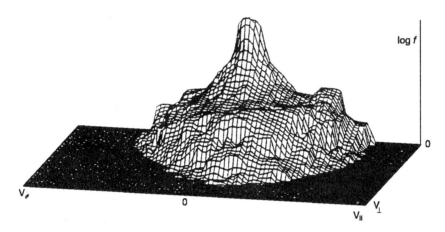

Figure 3.12. Mountain-like surface plot of an auroral electron distribution exhibiting a distinct beam at the edge of a relatively broad plateau. Parallel velocities range from -6×10^7 to 6×10^7 m s^{-1} and perpendicular velocities from zero to 6×10^7 m s^{-1}. The beam is centred on a parallel velocity of 2.7×10^7 m s^{-1} (2.1 keV). The central peak is limited in height by the lowest measured energy of 25 eV (from [104]).

3.2.3   At 'mid'-altitude

By 'mid-altitude', we mean altitudes which may be within the acceleration region, i.e. between 1000 km and 10 000 km.

3.2.3.1   Upward-flowing electrons

Figure 3.13 shows energy distributions obtained at an altitude of 1400 km within an auroral electron stream at 67° latitude close to midnight [105]. These distributions introduce an entirely new factor. The two distributions are for electrons travelling almost directly upwards and almost directly downwards. The new factor is that the upgoing electrons have, over a broad range of energies below about 300 eV, much higher velocity-space densities than the downgoing ones. The corresponding angular distribution, of figure 3.14, shows the high velocity-space densities of the upgoing electrons to be confined to pitch angles close to 180°. The downgoing electrons were also field aligned on the second of the two samples of this angle. Both upward and downward streams were found to be sporadic, as the difference between measurements at opposite ends of the graph, taken just 18 s apart, confirms. Full details of the energy and angular distributions of these streams have been very recently revealed in high resolution by the Fast Auroral Snapshop (FAST) satellite, and have been found to occur adjacent to regions of downward-accelerated auroral electrons [106, 107]

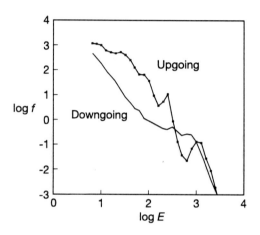

Figure 3.13. An example of a returning electron distribution exceeding, at low energies, its downgoing counterpart (from [105]).

3.2.3.2 Counter-streaming electrons

A feature differing from the above in degree rather than in kind is the phenomenon known as 'counter-streaming'. Figure 3.15 shows some measurements made at an altitude of around 8000 km from the satellite S3-3 [108]. The geomagnetic latitude was 67° and the time was one hour before midnight. Several cycles of a scan through the full range of pitch angles revealed a strongly field-aligned upward stream consistently exceeding its downward and also field-aligned counterpart. At higher energies than those shown in the figure, the upward streaming persisted, but the downward stream was highly sporadic. The FAST satellite shows that, at times, the counter-directed streams are almost exact replicas of each other to a very high degree of accuracy. Even when the

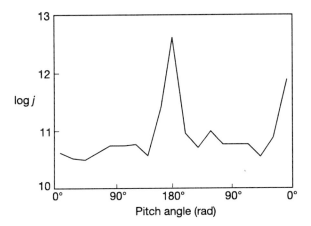

Figure 3.14. Pitch-angle distribution for electrons of 5 eV–15 keV for the event depicted in figure 3.13, revealing the strong field alignment in both upward and downward directions (from [105]).

downward stream is restricted to lower energies, these electrons still mirror very closely their counterparts in the upward stream [106].

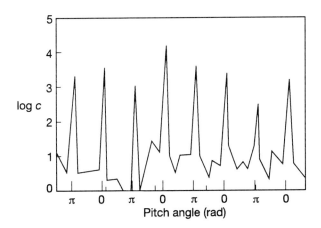

Figure 3.15. 'Counter-streaming' electrons. The count rate of electrons of 70–240 keV exhibiting streaming simultaneously parallel and anti-parallel to the magnetic field (from [108]).

The upward streams and counter-streaming, although not our prime concern, are found right in the heart of the acceleration region and so demand to be accommodated within any theory of this region.

3.2.4   At 'high' altitude

'High' altitudes in the present context are those above 10 000 km, i.e. generally above the acceleration region.

One of the difficulties in establishing that measurements were made above an acceleration region is that the electrons at this point can show no sign that they are about to be accelerated. However, it is sometimes possible to confirm that acceleration is occurring below by photographs taken from the same or other space vehicle or from the ground. Other indications will arise in later discussion of the wave model of acceleration.

3.2.4.1   'Conics'

A velocity distribution typical of those at altitudes above the acceleration region is shown in figure 3.16. This was obtained from measurements by the DE-1 spacecraft at an altitude of 11 000 km over the polar cap. Although this was not actually within the auroral zone, the authors confirm that the features observed here are also characteristic of the auroral zone. Of particular interest are the ridges or bulges of velocity-space density seen in the upward-bound electrons centred on pitch angles of about 135°, just outside the loss cone. This now commonly observed feature is known as a 'conic', a name stemming from the fact that the high velocity-space densities lie on a thick conical shell of directions. The half-angle of this conic is about 45°. The crescent-shaped ridge in the downward direction indicates that there has been some acceleration above the altitude of measurement.

3.2.4.2   'Widened loss cone'

A closely related feature also apparent in figure 3.16 is the seemingly widened loss cone at the higher energies. There is certainly no sharp boundary demarking a transition from particles of magnetospheric origin to particles of atmospheric origin at the nominal position of the loss cone. It is worthy of note that at the lowest observed velocities the distribution is almost isotropic and there is no evidence of a loss cone there at all.

Widened loss cones and conics are conveniently captured together in a spectrogram of count rate as a function of energy and pitch angle. Figure 3.17 is such a display based on measurements made from the Viking spacecraft at an altitude of about 12 000 km at a local time near noon. The vertical stripes at energies centred on 1 keV on either side of the loss cone centered on 180° are manifestations of a conic, while the widening of the pitch angle range of low count rates towards lower energies could be interpreted as a widening of the loss cone at these low energies. This figure epitomises the character of electrons above the acceleration region.

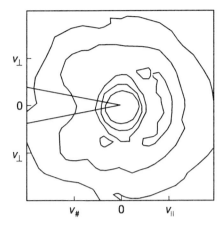

Figure 3.16. Electron distribution with a crescent-shaped peak in the $v_{\parallel}$ (downward) direction and a 'conic' just outside the loss cone in the $v_{\#}$ (upward) direction. Parallel and perpendicular velocities range from zero to 1.39×10^7 m s^{-1} (0–550 eV). Contours of velocity-space density are spaced at factors of approximately 16, the innermost being 6×10^4 km^{-6} s^3 and the outermost 1 km^{-6} s^3 (from [109]; see also [104]).

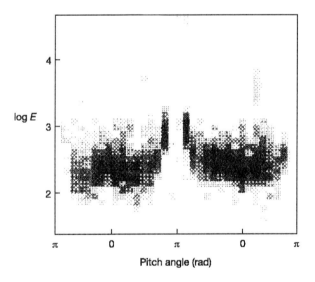

Figure 3.17. An energy–pitch-angle spectrogram (higher count rates shown darker) illustrating an electron conic (the near-vertical dark stripes on either side of pitch angles of π) and an apparently widened loss cone (the region of low count rate centred on π becoming wider towards lower energy). The measurements were obtained over two 20 s spin periods of the Viking satellite at an altitude of 10 000 km on 5 May 1986 (from [110]).

3.2.4.3 'Hollow'

Figure 3.18 contains another new feature. This is a slight field-aligned hollow at velocities just below a crescent-shaped peak. While this is not a frequently reported feature, it is one that has occurred on other occasions too, and so may offer an important clue to the mechanism shaping these distributions.

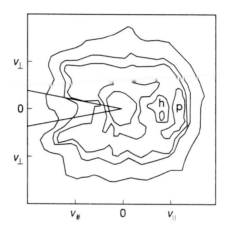

Figure 3.18. A crescent-shaped peak (p) and a 'conic' with the addition of a field-aligned hollow (h). Velocities range from zero to approximately 4×10^7 m s^{-1} (0–4.6 keV). The inner contours of velocity-space density are spaced at factors of approximately 1.8 and range from an innermost value of 30 km^{-6} s^3 to 5.6 km^{-6} s^3. The outermost contour is at 1 km^{-6} s^3. The peak attains more than 18 km^{-6} s^3 while the hollow dips to below 5.6 km^{-6} s^3 (from [109]; see also [104]).

3.3   The potential-difference theory

3.3.1   Beginnings

More than 20 years before direct measurements were possible it was suggested by Alfvén that the protons and electrons of a (then only postulated) solar wind were deflected in the geomagnetic field to produce two regions of space charge, a positive one on the dayside of the Earth and a negative one on the nightside [111]. Protons subsequently repelled from the dayside and electrons repelled from the nightside would, according to this suggestion, be guided along lines of force to the high-latitude upper atmosphere where they would produce the aurorae.

While admiring the imaginative perception that led to this very early suggestion of an enclave, being created in the solar wind by the obstacle of the geomagnetic field, and of regions of space charge being formed within this region, the reader may be somewhat puzzled by the mechanics of the proposed acceleration process. We found in Chapter 1 that acceleration requires temporal change, and, in particular, that the electric fields of static space charge, being conservative, do not cause (net) acceleration. We are, then, faced with a major conflict. It must either be the case that our conclusions in Chapter 1 are wrong or incomplete, or that the proposed accelerator simply will not work. Since a resolution of this conflict is crucial to our whole study, I would encourage you to examine the Chapter 1 conclusions again very carefully. If you are satisfied that static space charge cannot act as an accelerator, you will be ready to identify the flaw in this mechanism.

The flaw is that all kinetic energy gained by a charged particle on being repelled from one of the space-charge regions will have been expended on gaining entry to the space-charge region. This inescapable property of static electric fields and indeed all central forces may have been what Cowling was referring to when he proclaimed there to be 'a serious energy difficulty' with this mechanism [112].

Cowling's reminder might well have been the end of the story for this idea, but for an extremely interesting and, as we shall see later, far-reaching experiment in Alfvén's Stockholm laboratory. This showed that electrons were accelerated in an electrical discharge when an electrostatic double layer could be induced to form within the tube. The disabling conservative nature of the double layer (see Chapter 1) seems to have been well appreciated by Alfvén since his definition of this entity became 'the simplest configuration of space charge giving an electric field inside but with no electric field outside' [8]. No-one has ever suggested what this configuration, able to defy the fundamental conservatism of central forces, might be. Analyses were made of the internal electric fields of double layers in one dimension, with the inevitable and inescapable external fields of real three-dimensional double layers being left out of the reckoning. A belief arose from this, and continues to be held in some quarters that double layers would produce or 'support' a net change of electric potential and that this 'potential drop' would serve to accelerate charged particles (e.g. [113], responding to our claim [114] that this is fundamentally impossible).

Note, though, that the fact that the two phenomena, electron acceleration and a double layer, were correlated does not prove cause and effect. The possibility that, instead, they are both related to a third phenomenon will be explored towards the end of the chapter.

Nevertheless, an acceleration theory involving potential differences developed, and, with an interest in seeing how this could have happened, and to become acquainted with several new features of the acceleration region arising from the many and varied investigations prompted by it, we shall trace the development now.

3.3.2 Further impetus

Considerable impetus was given to the potential-difference theory by the first measurements of the electrons responsible for an auroral arc. McIlwain deduced, using a scintillator/photomultiplier in combination with a magnetic analyser flown on a rocket reaching an altitude of only 120 km, that the energy spectrum exhibited a sharp peak [115]. A concentration into a narrow range of energies could not, it was thought then, be caused by a stochastic or random process since random processes would always lead to a broadening of a velocity distribution, never a sharpening, just as with the tray of sand and the child's bricks. The sharp peak was thus thought to be the work of an ordered process. An ordered process was, of course, exactly what had been forecast for the potential-difference theory which saw the electrons being accelerated into a potential well.

The first direct measurements of the peak were made by Albert [116] in 1967 using a scintillator with an electrostatic analyser in front, flown on a rocket reaching an altitude of 250 km. The result, which was then quite startling, is shown in figure 3.19. This energy spectrum serves immediately to corroborate the sharp peak that had been deduced, but not directly seen, by McIlwain. In fact the peak was so sharp as to attract the epithet 'monoenergetic' for the spectrum as a whole. Even if the results are converted to velocity-space density and plotted on a logarithmic rather than linear scale, as in figure 3.20, the peak is still very apparent. The term 'monoenergetic' though seems somewhat less warranted from this perspective. You will have noticed that a key feature missing from figures 3.19 and 3.20 is the population of electrons now known to exist at low energies. This was because the scintillator/photomultiplier combination, though the most sensitive instrument available at the time, was incapable of detecting the low-energy component. This component awaited the arrival on the scene of the channel electron multiplier. Unfortunately, although the low energies were revealed by the more sensitive detectors [117], they were not immediately apparent on the count-rate distributions that were at first presented. The term 'monoenergetic' gained a firm hold, and survives to this day despite the wide range of energies now known to be involved.

Since near-monoenergetic electron distributions are just the type to be produced by acceleration into a potential well, the potential-difference theory began to bask in the glow of this circumstantial evidence.

3.3.3 The paradigm

The essence of the potential-difference theory is depicted in figure 3.21. According to this theory electrons gain their energies by being accelerated downwards by the electric field corresponding to the potential contours. The contours imply—indeed require there to be–a region of negative space charge from which the electrons are accelerated. The model is therefore exactly the same in principle as Alfvén's original. The space charge is just located in a

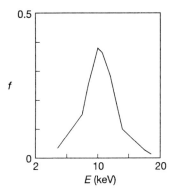

Figure 3.19. The first-reported auroral electron energy spectrum. Intensities, on a linear scale, exhibit a characteristic peak in intensity—on this occasion, close to 10 keV. The measurements were obtained from a Nike Tomahawk rocket flown from Fort Churchill, Canada on 16 September 1966 (from [116]).

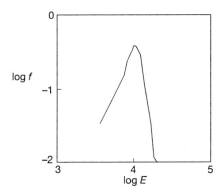

Figure 3.20. A logarithmic velocity-space density plot corresponding to figure 3.19, showing that, although there is still a pronounced peak, the distribution could not justifiably be termed 'monoenergetic'.

different place, i.e. much closer to the atmosphere, in recognition of the fact that the acceleration is now known to take place there. The model does not require the equipotentials to be precisely static—this would be inconsistent with the constantly changing pattern of the aurora—but they are considered to be quasi-static in the sense that they undergo little change while an electron traverses them. Since the time taken for an auroral electron to pass through the region is typically only a fraction of a second, and auroral arcs often remain stable for tens of minutes [87], so it is quite reasonable for an acceleration theory to be essentially a steady-state one. The reason the equipotentials curve upwards rather than downwards is that the latter can be ruled out by measurements of electric fields, which we shall look into shortly.

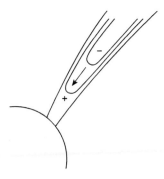

Figure 3.21. Essence of the potential-difference theory of auroral electron acceleration. Precipitating electrons (arrow) are accelerated by a quasi-static electric field. Note that, crucially, the equipotentials are open-ended lines (based on [118]; see also [119, 120]).

The model tacitly makes the major assumption that negative space charge has been marshalled into a configuration to produce the envisaged equipotentials and that there is some means of replenishing the electrons that are expelled. Reference is sometimes made to an 'auroral dynamo' which is able to accomplish this feat (i.e. to serve as the Van der Graaf generator's charging belt or the ski lift) but it is accepted even by the model's supporters that this vital aspect of the theory is unclear. Appeals have been made to a dynamo which combines the motion of the solar wind and the magnetic field of the Earth to give the charge separation. The closest attempt of which I am aware suggests how the equipotentials might be closed in two dimensions [121], but still fails to close them in all three [122]. The mechanism which, in this model, enables solar-wind particles to overcome and maintain potential barriers much too high for them to surmount operates in the unseen third spatial dimension.

We are left with no option but to conclude that the potential-difference model, as it currently stands, lacks a vital element, i.e. the means by which the electrons are raised to the potential from which they are considered to be accelerated.

3.3.4   Predictions

Despite the missing link, the theory has held, and continues to hold an acceptance that has become almost unquestioned in the overwhelming majority of journal articles on the subject.   The great attraction is, I understand, that if the shortcomings are overlooked or seen as a separate, distant, or someone else's problem, many of the observations summarised above are readily and simply predicted. Moreover, there are some bonuses (a merit for any theory) in the form of predictions of other observed features. Some proponents have told me that they like it because it is simple just to add energies!

The main attraction is that the characteristic peak in the distribution is accounted for immediately. Electrons entering a region of more-positive potential Φ will increase their kinetic energy by $e\Phi$. Since velocity-space densities will remain unchanged, as Liouville's theorem advises us, the velocity distribution is simply shifted bodily to higher energies as we saw in Chapter 1, to leave no electrons with E below $e\Phi$. If the distribution outside the potential well falls off monotonically with energy, as is the case with a Maxwellian, kappa distribution or power law, a sharp peak will be created at the energy $E = e\Phi$ as discussed in Chapter 1 and, as shown in figure 3.22, the accelerated primary electrons provide a passingly good approximation to the 'monoenergetic' distribution of figure 3.20 straight away.

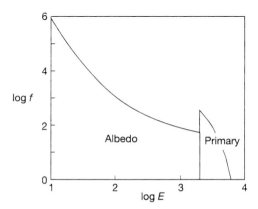

Figure 3.22. Interpretation advanced (see [123]) to account for the component of the auroral electron energy distribution below the peak. Following acceleration of primary electrons into the atmosphere, albedo electrons returning from the atmosphere are reflected back down by the potential barrier assumed to have accelerated the primaries.

The discovery of the electrons with energies below the peak in the distribution caused for a while a major bar to progress since the above considerations suggested that they should not be there at all. Evans, however, rescued the model from this apparent impasse by pointing out that the albedo of secondary and backscattered electrons rising from the atmosphere with energies too low to climb the potential barrier and escape to the magnetosphere would be reflected down again towards the atmosphere [123]. These particles would have just the right energies $E < e\Phi$ to occupy the portion of the distribution below the peak. Evans made a detailed estimate of the albedo to take this consequence of the model into account. The result appears schematically in figure 3.22. The proposition that the downward-travelling low-energy electrons were one and the same as the upward-travelling ones gained much support from measurements such as that seen earlier (in figure 3.9) where, indeed, the upward and downward low-energy electrons have the same velocity-space densities at low energies. In order to account for the fact that the observed distributions did not show the predicted sharp rise in velocity-space density between the highest-energy reflected albedo and the lowest-energy accelerated primaries, it was proposed, in accordance with experience of the propagation of near-monoenergetic distributions in laboratory plasmas, that energy would become redistributed to cause the peak to become broadened. Some success was found in attempting to account for specific distributions along these lines [124], with the quality of fit being noticeably better for the slighter and weaker peaks [125].

A disquieting factor was, though, that upward and downward velocity-space densities were not always the same, as for example in figure 3.23, and that, even in instances when they were approximately the same, the low-energy supposed secondaries were rather insensitive, as we have seen, to changes in the primaries which, by this theory, generated them. Another problem was that the model failed to explain how broadening of a peak could proceed beyond the point where a (stable) plateau had been reached and then go on to generate a monotonic bulge, such as those seen in the auroral rays of figure 3.5.

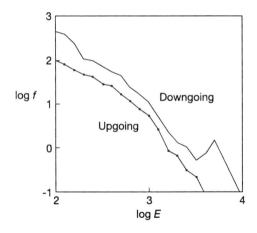

Figure 3.23. An observation of upward- and downward-flowing auroral electrons where the upward flux is less than the downward flux, even at energies below the peak, demonstrating that the downgoing electrons cannot be one and the same as the upgoing ones after reflection (from [126]).

The full implications of the potential-difference theory are perhaps best illustrated by means of contour plots of velocity-space density. Electron distributions to be expected for the straightforward case of a very localised accelerator are shown in figure 3.24 **A–D** at four different positions relative to the accelerator. **A** is far above the accelerator, **B** is immediately above, **C** is immediately below, and **D** far below, close to the atmosphere.

This investigation, although consisting of simple steps, is quite involved since there are many different routes to be followed before the full picture can be generated. This is by way of an apology for the intricacies of the analysis to follow.

We shall start at **A**, far above the accelerator, where the magnetic field strength is one half of that in the vicinity of the accelerator. At **A** the downward-moving electrons in sector **1** are from the source. Their velocity-space density falls monotonically with speed and energy, and they are isotropic over the downward hemisphere of directions. The electrons in sector **1'** are those members of **1** that have earlier been magnetically mirrored below and are on their way back to the source.

The downward hemisphere at **B** contains in sector **2** members of **1** that have small enough pitch angles (less than 45°) to penetrate to **B** without being mirrored by the doubled-in-strength magnetic field. Since there is no acceleration between **A** and **B**, velocity-space densities at all energies will be exactly as they were at **A**. Electrons in sector **2'** are earlier mirrored members of **2**.

At **C**, immediately below the accelerator, where the electrostatic potential is positive relative to the source, electrons of sector **3** are electrons from **2** accelerated by the potential difference Φ. This causes energies to be increased from E to $E + e\Phi$. This increase is irrespective of the electron's pitch angle. Since the accelerating force (the electric field multiplied by the particle's charge) is directed parallel to the magnetic field, the acceleration increases the parallel velocity $v_\parallel$ while leaving the perpendicular velocity $v_\perp$ unchanged. The amount by which the parallel velocity increases depends on the parallel velocity itself, and is such that

$$(v_\parallel^2)_C = (v_\parallel^2)_B + v_\Phi^2$$

where

$$v_\Phi^2 \equiv 2e\Phi/m.$$

The overall effect is for sector **2** electrons to shift towards higher $v_\parallel$ in such a way that contours of velocity-space density change from semicircles about the origin in **2** to arcs of circles, still about the origin in **3**, but with reduced spacing most noticeable at parallel velocities close to v_Φ. The highest velocity-space densities in **3** occur at small pitch angles, thus producing field alignment. Electrons in sector **3'** are earlier mirrored members of **3**.

We now move without further acceleration to **D** where the potential is the same as at **C** but where the magnetic field strength is 2.2 times as strong as at **C**. Electrons in sector **4**, which all have energies greater than $e\Phi$, are electrons from **3** having had their pitch angles increased by the magnetic field. The higher energies have undergone the greater increase because their pitch angles were greater at **C**. At the highest energies the pitch angles have increased to 90°, and some cases beyond to result in the mirrored component **4'**. If the electrons of **4** could travel below **D** into an indefinitely increasing magnetic field, they would all eventually mirror, so making the distribution at energies above $e\Phi$ completely isotropic, the contours becoming complete circles. However this

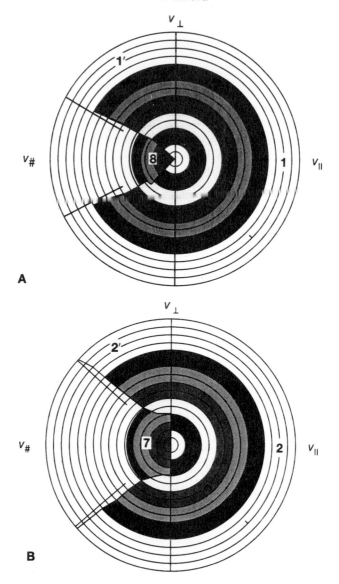

Figure 3.24. Evolution of the electron distribution as envisaged in the potential-difference theory of acceleration.   Distributions are shown for: **A**, a position far above the acceleration region; **B**, immediately above the accelerator; **C**, immediately below; and **D**, close to the atmosphere far below. The various subpopulations are identified by the attached numbers. The shading helps to trace the movement of the electrons of given velocity-space densities, which are conserved throughout. The evolution is discussed in detail in the text.

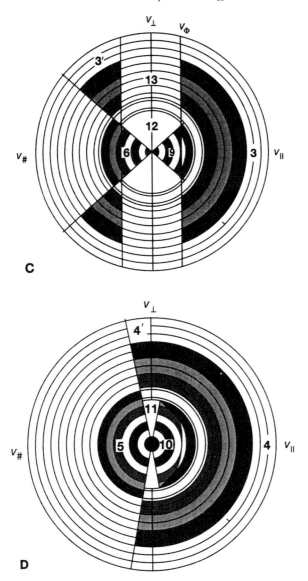

Figure 3.24. Continued

does not happen in practice because of the barrier to progress presented by the atmosphere. We represent the atmosphere here as a sharp barrier situated at an altitude where the magnetic field strength is just a little greater ($\times 1.13$, so that **D** corresponds to a typical 'low altitude'. An albedo of secondary and backscattered electrons emerging from the atmosphere as a result of the bombardment from above constitutes a family of secondaries **5** setting off upward (within the upward loss cone as indicated) to occupy sector **6** at **C**.

Those members of **6** with parallel velocities in excess of v_Φ will overcome what is now a potential barrier to reach **B**. Their parallel velocity will, in the inverse process to that above, be reduced according to

$$(v_\#^2)_B = (v_\#^2)_C - v_\Phi^2.$$

At **B** these electrons occupy sector **7**. As the result of the retardation, sector **7** is at low velocities considerably wider than the indicated nominal upward loss cone. These electrons continue upward to sector **8** at **A** where, again, they extend outside the nominal loss cone at low velocities.

Those members of **6** with parallel velocities below v_Φ are reflected by the potential barrier to occupy area **9**. These electrons then proceed down to become **10** at **D**, where those with pitch angles less than 70° enter the atmosphere to generate further albedo, i.e. further members of their own family. There is thus a positive feedback which needs to be approached in a detailed computation by a method of successive approximations evolving towards a self-consistent precipitating flux **4 + 10**, and secondary effect **10** [123].

While we have now seen how most of the components of the distribution at **D** arise, we have not managed to find any customers for area **11**. This is because this region of velocity space is inaccessible (in dynamical trajectories) from above or below. This 'forbidden zone' is also seen as **12** at **C**, where it covers a wider range of pitch angles, and extends also to higher perpendicular velocities in **13**. By the same token that particles are unable to enter from outside, any particles within will be unable to escape. In the radiation zones particles are trapped between magnetic mirrors. Here they would be trapped between a magnetic mirror below and a potential barrier above. The population to be expected is unclear. If no electrons were trapped when the conditions were established, the region of velocity space would remain empty. If non-dynamical trajectories were initiated by, for example, a scattering of directions by irregularities in the magnetic field, the 'forbidden' region might become filled with anything up to the same intensities as in **10**, or possibly, if there were energy changes as well, up to those of area **4**.

We should note at this point that if our accelerator were located just above **D** rather than between **B** and **C**, the distribution of downgoing electrons at **D** would be the same as that shown in **C**. The upward component would be different, though, since it would reflect the closer proximity of the atmosphere, and thus exhibit the same loss cone as at present. If the potential difference were distributed uniformly between **B** and **D**, the distributions within and below the accelerator would be intermediate between those in **C** and **D**.

The four parts of figure 3.24 summarise the basic effects to be anticipated from the potential-difference theory. We can now see how well these match up to the observations.

3.3.5 Tests

The most noticeable and most well known feature of the auroral electron distribution, the characteristic peak is certainly produced. It is manifest as the transition between components **10** and **4** in **D**.

Magnetic-field alignment is readily accounted for by the theory in the concentration of high velocity-space densities around small pitch angles at the low-energy boundary of **4**. If the electrons constituting the original peak were re-distributed into the lower-energy region they would help to account for the field alignment below the peak as well as above. Whether it could cause the field alignment to be more marked below the peak than above, as in figure 3.6, is unclear. If the accelerator were located at a high enough altitude, magnetic mirroring could be expected to draw the peak into a ring or ridge. Again, whether nose-like extensions along $v_{\parallel}$ like those of figure 3.10 would be generated remains to be discovered.

The apparently widened loss cone exhibited by the electrons **7** and **8** goes some way toward mimicking this feature prominent in figures 3.16 and 3.18. Differences in detail are apparent though, particularly for small values of $v_{\parallel}$.

The 'hollow' seen in figure 3.18 does not arise automatically. However, if it is taken that the forbidden region is well populated (one of the possibilities foreseen above) velocity-space densities of the electrons in **11** might exceed those in **10**, thus leaving a hollow in the right place.

The observation that the high-energy tail broadens does not fit, since the model's transition between the source, **1**, and precipitation, **4**, amounts to a simple addition of the same quantity of energy to all electrons, and this does not broaden the distribution. This shortcoming has long been recognised, and has been 'resolved' qualitatively, as mentioned above, by modifying the theory to include a redistribution of energy among the accelerated electrons, the effect of which is seen as tending to flatten the peak and broaden the whole distribution.

The common identity of upward- and downward-flowing low-energy electrons in this model is seen here in wider perspective. There is some support for this proposition in the near symmetry of the contours at the centres of figures 3.10 and 3.11. At least there is no obvious discrepancy. We have to bear in mind, though, that the upward and downward distributions do not always match, as we have recently seen in figure 3.23.

You may decide to set little store by discrepancies of this kind, attributing them to the aforementioned redistribution of energy, or perhaps to temporal fluctuations occurring while the measurements were being accumulated. On the other hand, they might begin to ring alarm bells.

Multiple peaks, the lower-energy one being field aligned the other not, as in figure 3.8, are not anticipated. In whatever manner the potential difference is distributed, all electrons travelling from **A** to **D** would gain the same energy. This limitation of the theory is well recognised. The possibility of a second source of electrons part way through the acceleration region has been mooted, but rejected on the grounds that no conceivable source has yet been identified. A suggestion that has been advanced to accommodate the lower-energy peak is that it is formed by a concurrent but totally different process (involving wave–particle interactions) to that producing the peak at the higher energy.

This theory now has waves playing important roles above the peak and below the peak, but not in the formation of the peak.

The upward streams of electrons seen earlier in figure 3.13 are sometimes interpreted, in a reverse polarity version of the potential-difference theory, as electrons from the ionosphere being accelerated into a region of positive space charge. This interpretation has received some support from the FAST satellite where changes in electrostatic potential evaluated from electric field measurements match within a factor of two the energies by which the electrons appear to have been accelerated. The electron energy distributions are, however, rather broader than acceleration into a potential well would produce, suggesting that wave–particle interactions also play a part. Counter-streaming is accommodated into this line of thinking by the invocation of barriers to progress in the form of negative potential wells located beyond the positive potential well [107]. Reflection by a negative potential well has the great merit of accounting very neatly for the close correspondence of the upward and downward elements of counter-streaming electrons of the same energy. Note, however, that the process responsible for upward, and accompanying downward streams has relevance to the question of the downward acceleration of auroral electrons only through the possibility of the two being assumed [120] to be mirror images of each other, and the vital question of the closure of the electrostatic equipotential surfaces around the potential wells being sidestepped.

A crucial test that presents itself, but which as far as I know has never been performed, is to confirm from simultaneous high- and low-altitude measurements from DE-1 and DE-2 that the electrons within the loss cone at high altitude (those of sector **8** at **A**) are duly attenuated versions of those in sector **5** at **D**, and that the attenuation corresponds to the energy of the peak in **4**.

The 'conic' is not predicted. Within the context of the potential-difference theory, the conic is seen as the product of yet another concurrent but independent process. The process envisages electrons having their perpendicular velocities increased in a cyclotron-like acceleration [109] driven by electron-cyclotron waves or in a more complex process involving ion-cyclotron waves [127].

In summary, then, the potential-difference theory does not perform too badly against the observations. It anticipates the principal features of the peak, the low-energy component and the field alignment below the accelerator. It also anticipates a wider than nominal loss cone above the accelerator. With added processes it can accommodate a broadening of the high-energy tail, and field alignment of the low-energy electrons. There are several discrepancies, though. Some can be explained away, but others such as multiple peaks, counter-streaming electrons and conics clearly require concurrent separate mechanisms. Having come so far, it does seem at least to be worth looking into what evidence there is for the electric fields upon which this theory is based. In any case, a significant electric field would have implications for any theory.

3.3.6   Search for parallel-to-B electric fields

3.3.6.1   *Direct measurements*

There has been a concerted search for the electric fields invoked by the potential-difference theory for several decades and employing many different techniques. Expectations range from a uniform field of ~ 1 mV m^{-1} over distances of some 10 000 km, to much stronger, spatially confined fields. Neither of these extremes would, it has always been recognised, be easy to observe. Weak parallel fields could easily be lost in confusion with the larger perpendicular field and fields induced by spacecraft motion through the geomagnetic field, while the very localisation of strong fields would make encounters unlikely.

Probes carried on spacecraft to measure potential differences between the ends of long booms have so far yielded no direct indication of a parallel electric field. In fact it has recently become clear that there is on average no extensive field greater than 0.1 mV m^{-1}, with the measurements being consistent with zero [128]. Fluctuations in potential that have been seen have been heralded as electrostatic double layers [129]. Potential differences across most of them were confirmed to be zero, but some of the fluctuations were estimated to have net potential differences of around 0.1 V [130]. The net voltage was interpreted, as a discovery of Alfvén's postulated double layers, without comment on the possibility that net voltages occurring in a few cases might have resulted from experimental error rather than from a breakdown of fundamental physics (see [131]). The interpretation goes on to consider that the potential differences of many such double layers would add together to provide the substantial potential differences required by the theory. If we accept for a moment that the net potential differences were real, and they would add together like battery cells wired in series, it is immediately apparent that the number required to generate 10 kV would be around 100 000. Their average spacing would then, if the acceleration is to be accomplished in the typical distance of 10 000 km, be around 0.1 m. But 0.1 V in 0.1 m represents an average electric field of 1 V m^{-1}, which is much greater than that observed.

The proposition that was earlier found to be irreconcilable with fundamental physics seems in addition, therefore, to conflict also with observation.

3.3.6.2   Ion tracers

Numerous and various attempts have been made to reveal the supposed parallel electric field indirectly. One approach was to use rockets to inject into the acceleration region clouds of barium atoms to be ionised by the action of sunlight to deposit ions which would be accelerated upwards by the electric field [132]. The ions are created in an excited state, recovery from which involves the emission of light. This emission allows events to be followed visually from the ground. The technical and logistic difficulty of such experiments speaks volumes for the conviction with which the belief in the field was, and is still, held and for the determination to demonstrate its presence.

In one of many such attempts [133] the envelope of light emission was seen to accelerate upwards from an altitude of 7500 km when auroral arcs appeared in the general area, while a neighbouring ion cloud did nothing. The apparent acceleration corresponded to a substantial energy gain of several keV. On other occasions, though, upward acceleration has been seen or inferred without an accompanying aurora, and clouds above aurorae have not been accelerated. The lack of a consistent link between injected ion motion and the aurora leaves little choice but to conclude that any quasi-static parallel electric fields that do exist or occur in the acceleration region are unimportant in the electron acceleration process. Note also that, in any case, the envelope of emission does not necessarily trace the progress of individual ions, since ions at different locations could conceivably become excited and subsequently emit radiation at different times, just as a Mexican wave travels around a stadium without the spectators needing to do so!

3.3.6.3   Electron tracers

In an equally complex and difficult series of experiments, electrons have been fired from electron guns carried on rockets in attempts to sense remotely the barrier to upward progress presented by the downward-accelerating potential difference [134]. These experiments sought, in effect, to reproduce the reflection of secondary and backscattered electrons invoked by Evans. Using a pulse-coded beam with a number of different energies, and employing several electron sensors to increase the chance of intercepting and identifying returning electrons, reflections have been detected. However the unpredictable energy dependence did not match up to a simple interpretation and it was concluded that the process causing the echoes was not as straightforward as originally envisaged.

The attempts, then, to detect the envisaged parallel electric field, either directly as a potential difference or through its effect on injected ions or electrons have yielded nothing that could be considered at all compelling. Numerous attempts have been made though to detect the fields indirectly.

3.3.6.4 *Indications from transverse fields*

Electric fields are readily measurable transverse to the magnetic field. Magnitudes of tens or even hundreds of mV m^{-1} have been detected quite routinely in and around the acceleration region [135]. Just such fields are expected by the potential-difference theory, as clearly implied by figure 3.21. In support of the view that the transverse fields corresponded to those of the potential-difference model, the fields were indeed directed inwards, and, when summed over the appropriate distances corresponded to several kilovolts, again more or less as required as long as it could be assumed that there were still greater potential variations at higher altitude. There was even the circumstantial evidence that potential differences obtained from electric fields measured at high altitude were greater than those at low altitude [136]. Along with this was the consistent finding that electric fields close to the atmosphere were insubstantial, showing that, if the equipotentials did cross magnetic lines of force, they curved upward as shown. This was, though, evidence for the faithful rather than for sceptical.

3.3.6.5 *Indications from precipitating positive ions*

It was natural enough to seek evidence of the parallel electric field in the behaviour of positive ions, mostly protons, that generally accompany electron precipitation. Electric fields that accelerated electrons downwards would naturally retard ions moving in the same direction and accelerate upgoing ones. Such behaviour, occurring consistently, as indeed it did [137], would have been strong evidence, if not for a net potential difference, at least for a substantial parallel electric field.

As soon as detectors capable of measuring accurately the distribution function of the ions were flown into discrete forms of aurora, it was clear that there was no consistent anti-correlation between downward-flowing electrons and ions [100]. Perhaps, however, a judgment based on this would have been too harsh. After all, there was no reason to believe that the changes noted in the electron and ion distributions were due solely to changes in the accelerator. The source distributions may well have been different at different locations, serving to confuse the issue. Maybe.

3.3.6.6 *Indications from upward-flowing ions*

A prediction that did meet expectation, however, was that, on field lines where electrons were accelerated downwards, positive ions were accelerated upwards, or at least could be interpreted this way [137]. This result constituted for the adherents of the theory an unambiguous proof, and remains today a mainstay of the potential-difference theory.

A direct quantitative comparison between electron and ion acceleration was not possible at first. Electrons had reached the observing satellites from above, and ions had reached them from below, thus having traversed different regions of space. It was not even totally clear that the ions' upward acceleration was magnetically conjugate to the electron acceleration. Tests we have performed on measurements from the Viking spacecraft [138] suggest to us that ions attain their final energies below the altitudes where electrons are accelerated [139].

Complications arose. The upward-flowing ions could be categorised into ion beams, whose distribution peaked in the forward (upward) direction, and into ion conics, similar to electron conics and also peaked at pitch angles intermediate between 90 and 180° [140]. Ion conics were, and continue to be, seen as products of an acceleration preferentially perpendicular to the magnetic field caused by ion-cyclotron waves. The ion 'beams', though, were presumed to be the direct result of an upward-directed electric field. The distinction between 'beams' and 'conics' could be made only after the exercise of great care, since the mirror force caused the opening angle of the conic to close as altitude was gained, allowing conics formed at low altitude to masquerade as beams at high altitude.

The recent high-resolution measurements from the FAST satellite, within what appears to be the lower part of the acceleration region itself, have considerably clarified the picture. Here, beams and conics are easily distinguishable. They are found to be mutually exclusive, yet both lie within regions of accelerated electron precipitation. A key finding is that the upward travelling ions are found consistently to lie within negative potential wells. Moreover, the depth of the potential wells corresponds with the energy by which the ions appear to have been accelerated [107, 141]. This match requires arbitrary assumptions to be made, though, about the ionospheric potentials from which ions in different parts of the beam originated. The tacit assumption is also made that, despite the generally oscillatory nature of the electric field, variations in the smoothed field represent primarily spatial structure rather than temporal fluctuations. Even if one is unable to endorse all of these assumptions, the fact that the ion beams exist within negative potential wells seems to be established by these results. Some confirmation is provided by an equivalent reduction in the energy of downward-accelerated auroral electrons found in these regions. Whether one can go so far as to conclude from events surrounding the upward acceleration of ions that there is no doubt left that 'parallel potential drops' are a dominant source of auroral particle acceleration [120] before the inevitable compensating 'potential jumps' which will slow the ions down again have been taken into account, before other contending acceleration processes have been eliminated, and before the complexities to be related below have been resolved, seems somewhat doubtful.

A key quantitative test that did present itself was the comparison of ions of different species that had traversed the same region of space in their upward flow. It was soon discovered that the mean energy attained by oxygen ions, assumed to be singly charged, was greater than attained by simultaneously observed protons [142]. The ratio of mean energies lay over the whole range 0.9 to 5.0. Relaxation of the assumption of single charge does not help to resolve what is to the potential-difference theory a major discrepancy. Adherents of the potential-difference theory offer an 'explanation' which has the oxygen ions gaining additional energy subsequently from another form of acceleration. Another way around the discrepancy is to assume that the different species of ion have arrived at the satellite from source regions in the ionosphere at different potentials. Only if we are prepared to accept a possible explanation as being equivalent to the removal of a discrepancy can we see the result of this test as lending much support to the potential-difference theory.

Recent results [143] have confirmed these findings and added further to the complexity by showing that helium ions, having intermediate mass, gain an intermediate quantity of energy. Moreover, the energies of hydrogen, helium and oxygen ions appear to vary in proportion to one another rather than by the same amount. These findings clearly introduce an extra complication into the match believed to exist between ion energies and depths of potential wells.

3.3.6.7 Comparison of upward-flowing ions and downward-flowing electrons

Yet another quantitative and potentially very revealing test was devised. The two DE spacecraft were used to compare the acceleration of downgoing electrons with that of upgoing ions flowing through the same region of space [74]. The prediction was, of course, that the two energy gains would be the same. They were not, but were close enough for mitigation (e.g. inexact conjunctions) to be found for the differences. Estimates of the potential difference gleaned from the widening of the electron loss cone returned estimates of potential difference in reasonable accord with the gains in proton energy. It has to be noted though that there were discrepancies amounting in places to a factor of five or more in energy gain, even after the two data sets had been 'adjusted' by introducing a time shift to give the best possible fit. These discrepancies could equally well be taken to imply that a potential difference between the high and low altitudes, if it exists at all, is too small to have any significant effect.

3.3.6.8 Upward ions and electrons

In another important comparison of electron and ion behaviour, it has been discovered that upward-directed electron beams are sometimes accompanied by ions which also show signs of having been accelerated [144] upwards. In order to account for this behaviour within the framework of the potential-difference model it has been suggested [145] that the phenomenon may be due to a time-varying electric field whose fluctuations appear static for electrons during the relatively short time they are within the accelerating region, accelerating them upward during part of the fluctuation cycle and downward in the complementary phase. The fluctuations average out, though, for the more slowly moving ions, leaving them subject only to the average, electric field. An agency responsible for the rapid ($\sim$ 1 Hz) repeated reversal of electrostatic potential between the ionosphere and the magnetosphere has not yet been identified. Neither have the implications of the invoked rapid changes of potential been evaluated. This puzzling issue may now have been resolved by the FAST satellite whose high resolution reveals that the electron and ion beams are found in neighbouring, but mutually exclusive, regions [107].

3.3.6.9 Summary

The many and varied attempts to detect the static electric field postulated to accelerate auroral electrons have revealed what appear to be potential wells, both positive and negative, above the auroral zone. The depths of the potential wells seen so far are considerably less than would correspond to the energies of accelerated auroral electrons. Their physical extent is unknown. There is no evidence for the net potential difference between a replenishable source and the auroral atmosphere (tacitly) required by the potential-difference theory.

3.4   The wave theory

Let us make a fresh start.

3.4.1   Beginnings

A consistent feature of the auroral acceleration region is its fluctuating electric field, e.g. [110, 146]. An example is shown in figure 3.25. Could these, or their like be responsible for the electron acceleration? Electric fields certainly exert a force—one essential element of an accelerator—and fluctuations, by their very nature, contribute the other essential element—temporal variation. Landau predicted in 1946 [9], significantly just after the potential difference theory emerged, that fluctuations of electric field and potential in the form of waves travelling through a plasma would cause a plateau to develop in an electron velocity distribution around the point where electron velocities matched wave velocities. The prediction is now confirmed to be correct, and the process, known as Landau damping (e.g. [21, 147, 148]), by which electrons gain energy from waves and the waves become weakened or damped down is one of the most important in the whole of plasma physics.

Acceleration by the travelling crests and troughs of electric potential is often likened to surfing. This analogy is helpful to bear in mind when looking into its actual workings. It is important to remember, though, that it is only the resonance aspect of surfing to which attention is drawn here: the many

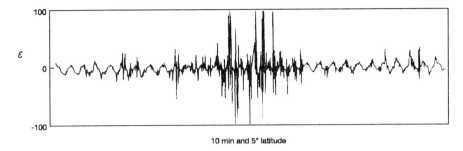

Figure 3.25. Electric-field fluctuations (in mV m^{-1}) recorded in conjunction with the electron conic of figure 3.17 (from [110]).

other factors coming into play in ocean surfing, such as viscous forces and active control, do not feature. The surfer, though, makes a convenient icon for the wave theory to contrast with the downhill skier of the potential difference theory.

Fluctuations like those of figure 3.25 arise when a plasma is disturbed by, for example, changes to its boundaries or the injection of new particles. Such disturbances excite trains of waves which propagate through the plasma like ripples on a pond, or sound waves through air. The situation in a plasma is, though, much more complicated than in these analogues. The long-range and collective influences give rise in a plasma to many types, or modes of oscillation, oscillating at different frequencies and propagating at different speeds. Two of these modes are of particular relevance to a plasma's electrons. The first is the electron cyclotron wave [149, 150], whose electric field is perpendicular to the magnetic field and rotates at the electron cyclotron frequency. We discussed earlier how this wave is capable of increasing the perpendicular velocities of gyrating electrons, as though they were in a cyclotron. The second is known by the awkward and off-putting name of the lower-hybrid wave.

3.4.2 Lower-hybrid waves

The lower-hybrid wave [149]–[153] is particularly adept at increasing electron velocities parallel to the magnetic field, a property which immediately raises its profile as a candidate for the auroral electron accelerator.

The mechanism of acceleration by lower-hybrid waves is not straightforward. It is complicated as are most engines, natural or manufactured, by being composed of many different working parts all making vital contributions. It also has to work randomly by chance. We saw at the beginning of this chapter that seemingly strong arguments were advanced from the outset suggesting that it was impossible for sharp peaks to be generated by random processes. The theory to be advanced below is based on the fact, demonstrated in Chapter 1 and in the statistical experiment invited by Appendix 3, that this conclusion, though very plausible, is not wholly correct.

3.4.2.1 Nature

A lower-hybrid wave consists of a series of crests and troughs in electrostatic potential propagating through a plasma The crests and troughs are caused by localised and temporary departures from exact balance of the charges carried by positive ions and electrons. These particles oscillate in sympathy with the electric fields striving to restore neutrality, and in so doing communicate the disturbance to neighbouring particles in a knock-on, or domino effect. Individual electrons and ions make no net progress with the passage of the wave. Since the wave is governed primarily by restoring forces that are essentially electrostatic and there is little or no associated magnetic field oscillation, the wave is described as an electrostatic wave. Note, though, that there nothing static about it.

The ambient magnetic field pervading the plasma plays a vital part. Without it, a disturbance to local charge balance or neutrality just continues to oscillate at the ion and electron plasma frequencies. The presence of a magnetic field promotes a hybrid coupling of electron oscillations parallel to the magnetic field with ion oscillations perpendicular to the field. The result is that the crests and troughs of the wave become almost, but not quite, aligned with the magnetic field, as figure 3.26 indicates. The misalignment is small but significant. The angle in radians is of the same order as the square root of the ratio of the masses of electrons and the plasma's ions. For an electron–proton plasma the angle is just over 1°. The significance of this is that parallel oscillating electrons and perpendicularly oscillating ions have approximately the same energies. A key property of the wave, therefore, is to couple, or convey energy between the perpendicular motion of ions and parallel motion of electrons of similar energy.

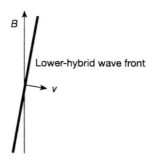

Figure 3.26. Orientation of a lower-hybrid wave front. The angle of inclination to the magnetic field (here exaggerated) is typically around 1°.

The crests and troughs, and indeed all intermediate phases of lower-hybrid waves, travel with a velocity component perpendicular to the magnetic field—perpendicular phase velocity—which is close to the typical random or thermal speed of the plasma's ions. This is usually much less than the electron thermal speed.

A wave such as this, which travels perpendicular to the magnetic field at a speed which is very much less than that of a plasma's electrons, may not at first seem a promising candidate for accelerating auroral electrons when we recall that, in order to meet Landau's criterion for resonance, we are looking for a wave travelling parallel to the magnetic field at a velocity comparable to that of the electrons. However, as we shall now see, the velocity with which the phase of the wave travels parallel to the magnetic field—the parallel phase velocity—meets this criterion perfectly.

The easiest way to appreciate how well parallel phase velocities match those of the electrons is to construct two set of parallel lines, as in figure 3.27, and to move one set (representing crests and troughs of potential) across the other set (representing magnetic lines of force) at a slight angle. The speed at which highs and lows of potential move parallel to the magnetic field will be revealed by the interference pattern of shadows to be very much higher than the motion normal to the magnetic field. The effect is exactly the same as that which causes crests of waves arriving nearly parallel to a sea wall to rush along the wall at a speed much higher than the incoming speed. In a plasma composed of protons and electrons, the parallel phase velocity of lower-hybrid waves is approximately 42 times the perpendicular phase velocity. Lower-hybrid waves are thus able to match at one and the same time the speeds of ions moving perpendicular to the magnetic field and electrons moving parallel to the field with the same kinetic energy.

3.4.2.2 Frequencies

Lower-hybrid waves exist only for frequencies at and above a characteristic frequency of a magnetised plasma: the lower-hybrid resonance frequency. The lower-hybrid (angular) frequency ω_{lh} is given by

$$\omega_{lh}^2 = \frac{\omega_{pi}^2}{1 + \omega_{pe}^2/\omega_{ce}^2}$$

where ω_{pi} is the ion plasma frequency, ω_{pe} is the electron plasma frequency and ω_{ce} the electron gyro-frequency. This expression shows that in a diffuse plasma with a strong magnetic field (i.e. in a plasma where the ion and electron gyro-frequencies greatly exceed their corresponding plasma frequencies)

$$(\omega_{lh})_{diffuse} = \omega_{pi}.$$

At the opposite end of the scale, a dense, weakly magnetised plasma has

$$(\omega_{lh})_{dense} = \sqrt{\omega_{ce}\omega_{ci}}.$$

The former extreme applies in the auroral acceleration region, while the latter applies at the bow shock and other regions of interest in later chapters.

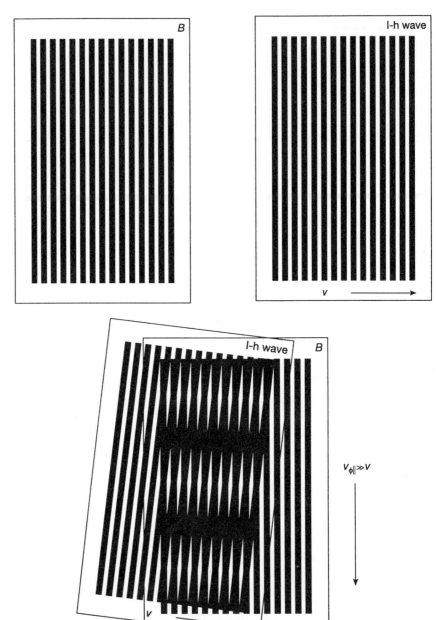

Figure 3.27. Demonstration by means of an an interference pattern that the phase velocity of lower-hybrid waves is much greater parallel to the magnetic field than perpendicular to the field.

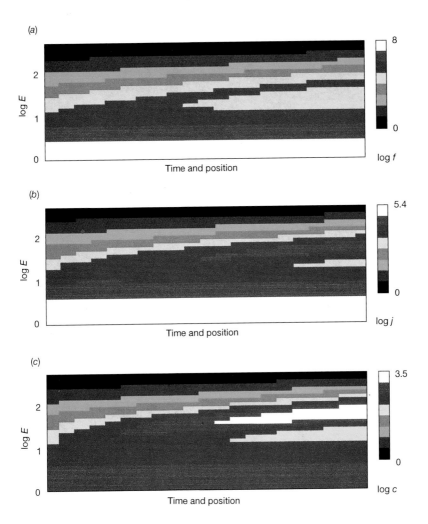

Plate 1. The colour-coded energy–time spectrogram is one of the most useful of presentation schemes. Its vertical columns indicate the energy dependence of the distribution at a given time, while the manner in which these vary with time and position trace the evolution. (*a*) shows the evolution of velocity-space density, f, in the case of a peak in this quantity emerging from an initial kappa distribution. The earlier presentation examples are 'snapshots' taken at the mid-point of the sequence. (*b*) shows the same information using differential intensity, j, as the colour-coded quantity. (*c*) uses count rate c of a differential analyser. It should be noted that the peak occurs at different energies in the three representations—appearing at the highest energy when seen in terms of count rate.

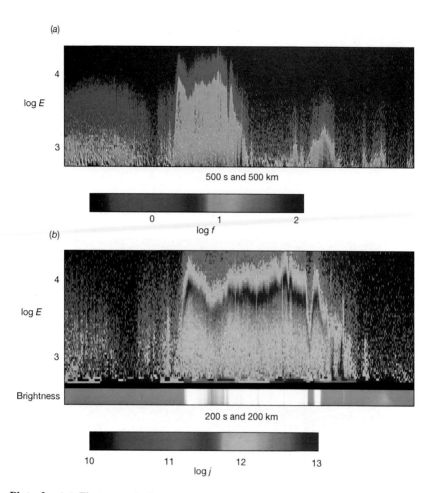

Plate 2. (*a*) Electron velocity-space densities observed from a Skylark rocket flight made from Andøya on 21 November 1976 through the stream of particles producing an auroral arc [100]. Over the first 100 s of measurement the monotonically falling distribution is reminiscent of the plasma sheet. This is followed by an intriguing and commonly observed dearth of electrons. The electron distribution within the arc stream, encountered next, varies from being slightly peaked to having a broad plateau. Poleward of the arc, the precipitation is weaker, less energetic and sporadic. (*b*) Electron intensities for the central section of plate 2(*a*). A sharp peak in the energy spectrum and its variable location in the energy band are evident from the persistent but erratically positioned red trace. The lower horizontal strip represents an attempt to evaluate from the energy flux carried into the atmosphere, the apparent brightness of the aurora, and in particular to discover whether the variations in energy flux would be perceived as significant variations in light intensity, thus giving the appearance of a multiple arc. It seems clear that if the arc was stable it would have been seen by an observer directly underneath as being a multiple arc.

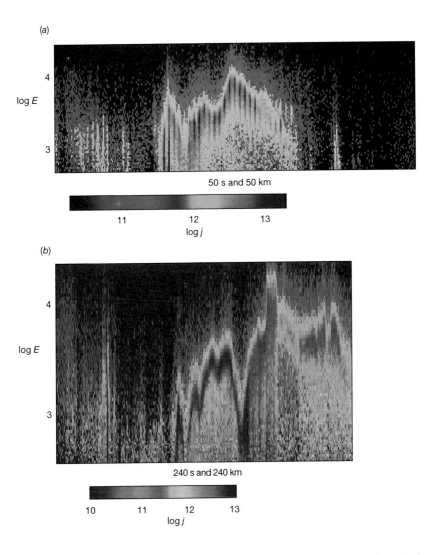

Plate 3. (*a*) Auroral arc electrons recorded on the flight of a Fulmar rocket from Andøya on 16 October 1977. The vertical stripes arise from the detector continuously spinning through a wide range of pitch angles in a highly field-aligned electron distribution. (*b*) Electrons in the vicinity of and within an auroral arc stream observed from a Nike Tomahawk rocket flown from Andøya on 27 November 1976. By great fortune, the arc developed a series of rays just as the rocket passed through the core of the electron stream. This lead to the brief episode of variable high intensities at energies above 10 keV. Energy distributions from this period appeared in figure 3.5 [76].

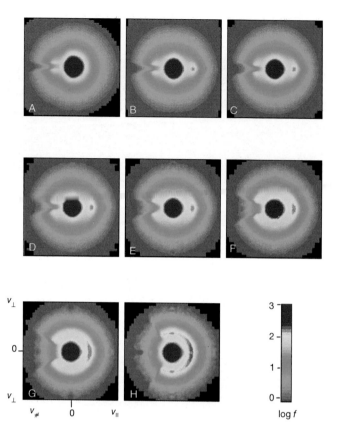

Plate 4. Evolution of an auroral electron energy distribution according to the wave model. The initial distribution is of power-law form, isotropic over the downward hemisphere. The resulting distribution is examined at eight altitudes A–H evenly-spaced between 24 000 km and 1000 km. For this illustration, acceleration takes place uniformly over the first 2000 km of the downward journey, i.e. immediately below A. Parallel ($\parallel$), anti-parallel (#) and perpendicular ($\perp$) velocities range from zero to 80 000 ms^{-1} in each panel. Electrons acquiring speeds greater than 80 000 ms^{-1} are not processed further, so the 'corners' of the panels should be disregarded. Many of the features found consistently in observed distributions are prominent. At A, the source ditribution is joined by a conic. Acceleration produces a field-aligned beam at B. The beam steadily widens to become crescent-shaped by G, and almost isotropic over the downward hemisphere at H. Albedo electrons fill a loss cone which is at its widest just above the atmosphere at H, and most narrow immediately inside the conic at A. A crucial feature of the wave theory, very apparent in this sequence, is the symmetry outside the loss cone about the zero parallel velocity axis, revealing that the conic is simply the magnetically mirrored outer part of the downgoing beam (adapted by C H Perry from [161]).

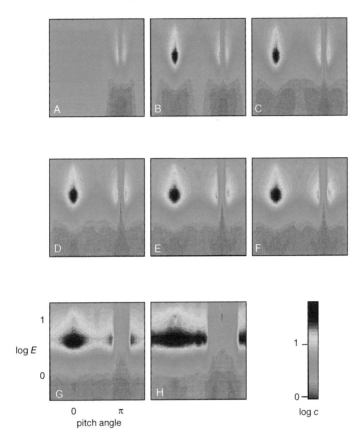

Plate 5. The same information as shown in plate 4, presented as the evolution of the energy–pitch-angle spectrogram, colour coded this time in terms of count rate. The conic appearing in A as two vertical yellow stripes at pitch angles close to π radians is well able to stand comparison with its experimental counterpart in the earlier figure 3.17. The apparent widening of the loss cone towards lower energies (indicated by the darker blue area) is due to electrons being drawn by acceleration from this source region to create the field-aligned beam. The heating produced by extension of the distribution to high energies is very apparent in this form of presentation (adapted by C H Perry from [161]).

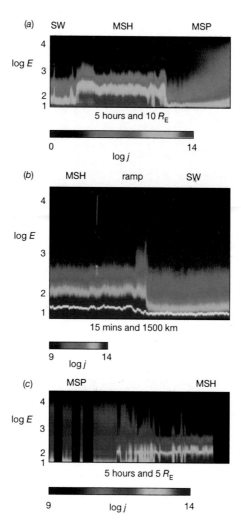

Plate 6. (*a*) Electron intensities along the inward leg of the UKS orbit on 4 November 1984 during which the spacecraft travelled from the solar wind (SW), through the magnetosheath (MSH) to the magnetosphere (MSP) [199]. (*b*) Electron intensities at the transition across the bow shock from the magnetosheath, via a 'ramp' in magnetic field strength to the solar wind, recorded from UKS on 6 October November 1984. This crossing is used for the case study. (*c*) Electron intensities observed from UKS during the crossing from the magnetosphere to the magnetosheath on 23 November 1984 at the location shown in figure 7.1. The crossing of the boundary layer was both prolonged and erratic, demonstrating the existence of a substantial and oscillating boundary layer. The three separate stretches of measurements near the beginning represent a 'real-time' search for the boundary layer while conserving power for the exploration of the layer itself [199] (data courtesy D S Hall).

(a)

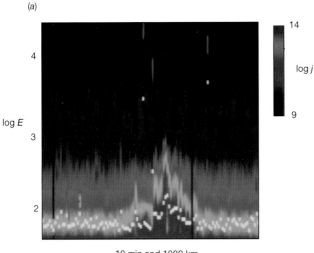

log E

14

log j

9

10 min and 1000 km

(b)

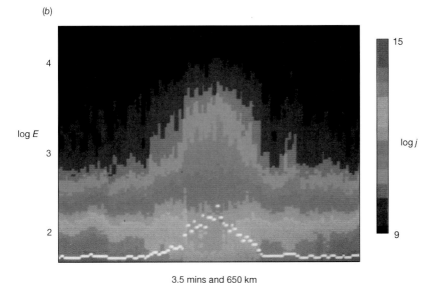

log E

15

log j

9

3.5 mins and 650 km

Plate 7. (a) Electron energy–time spectrogram of measurements by UKS on the occasion of the IRM release of barium into the solar wind on 27 December 1984 [194]. (b) Electron spectrogram for the 'solar wind event' of 28 October 1984 [197]. The white lines trace the development of the mean energy.

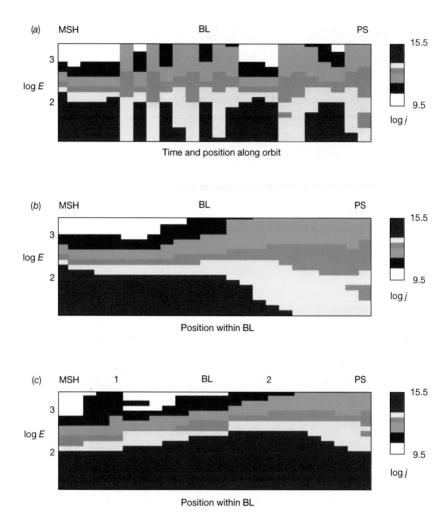

Plate 8. (*a*) Simplified version of 'perpendicular' electron energy–time spectrogram for the case-study magnetopause crossing. (*b*) As (*a*), with the columns of the spectrogram re-ordered to give a coherent picture. At low energies intensities fall smoothly from the solar wind to the plasma sheet, while at high energies the transition is in the opposite direction. At mid-energies (the energy of the common crossing point of figure 7.2) there is little change across the boundary layer. (*c*) 'Parallel' electrons re-ordered in accordance with the 'perpendicular' electrons. Part-way through the layer **2**, intensities at some medium energies exceed those at either end. There is also a 'non-conforming region' **1**.

3.4.2.3 Phase velocities

The phase velocity perpendicular to the magnetic field approximates to the sound speed in the medium, i.e.

$$v_{\phi\perp} \approx \sqrt{(3kT_i + kT_e)/M}$$

where M is the ion mass. The phase velocity parallel to the magnetic field depends on frequency, ω and is given by

$$v_{\phi\parallel} \approx v_{\phi\perp}\sqrt{\dfrac{M/m}{\omega^2/\omega_{lh}^2 - 1}}$$

m being the electron mass.

3.4.2.4 First application to the aurora

The first attempt, of which I am aware, to see whether lower-hybrid waves could, through these properties, serve as auroral electron accelerators was made by Swift in a paper published in 1970 [154]. Using also his earlier finding [151] that the parallel phase velocity of lower-hybrid waves would increase on a passage from high to low altitudes, he showed that, subject to the details of an encounter, some electrons would become caught up in waves and then carried forward in a surfing action. As the lower-hybrid wave accelerated, it would accelerate its passenger electrons with it. Swift concluded that, if such waves were present they could, indeed, lead to the almost monoenergetic beams then declared to exist. The process was, though, found to be very dependent on the wave remaining orderly and coherent, as in a linear accelerator. This seemed a tall order for the natural medium of the auroral acceleration region. As soon as what was considered to be a more realistic state of affairs was investigated by introducing random changes of phase, all that resulted was a slight broadening of the velocity distribution, just as indeed might be expected of an ill-adjusted linear accelerator.

Having just reviewed the velocity and energy distributions actually observed, you might see these results as being actually very promising. We ventured to suggest as much when attempting to account for one of our sets of measurements in 1978 [155]. However, as far as I am aware, there was no further development of this particular lower-hybrid wave model.

3.4.3 A wave model

The model we have subsequently developed is based on interactions taking place during random encounters between electrons and lower-hybrid waves. We accept from the outset that the waves will be less like those approaching a beach, but more like those creating the irregular turbulent chaos of a mid-ocean storm.

It will be appreciated at once that even the most powerful of today's computers would be hard put to follow the motion of a suitably representative sample of charged particles through a turbulent sea of constantly changing electric fields, even if the latter could be specified. Some exercises have been performed along these lines to indicate the types of effect to be expected [156]–[158], but none has proceeded far enough to be applied to the aurora.

We decided, therefore to simplify the problem. Our approach was to stylise both the turbulence and the response of electrons to it, while retaining the essential element of random encounters between the particles and the waves.

3.4.3.1 Wavepackets

The frequencies of lower-hybrid waves can, as specified above, lie anywhere between the lower-hybrid resonance frequency and a frequency of several times this value. In the auroral acceleration region the lower-hybrid resonance frequency is around 1 kHz, so the frequencies are about the same as those on a piano between two Cs above middle C and the top of the instrument. If all of these frequencies are produced at once, whether as lower-hybrid waves or sound waves, the result of course, is quite a cacophony. Not only do we experience these frequencies, but we also get new ones set up as beats between them where they reinforce or cancel out each other at regular or irregular intervals. Figure 3.28 shows how two simple oscillations combine to produce beats consisting of large-amplitude swings interspersed with relatively quiet spells. The pattern can be described as a series of wavepackets—in this simple case, in a regular train. If the component waves all have the same velocity, as sound waves do for example, the wavepackets retain their identity and travel at this common velocity. The situation is more complicated if velocity depends on frequency, as it does for lower-hybrid waves.

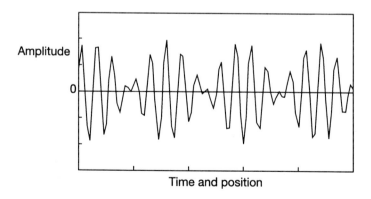

Figure 3.28. A sequence of beats or wavepackets resulting from superposition of two waves of differing frequency.

As more frequencies are introduced the appearance and disappearance of wavepackets becomes more and more irregular, arriving ultimately at the chaotic result akin to the turbulent ocean. To represent this likely situation, we consider the medium to be pervaded by numbers of localised, short-lived wavepackets.

3.4.3.2   Interaction between electrons and wavepackets [23, 159]

We consider now how electrons travelling through the region might be affected by wavepackets that they encounter. For this we return to the type of kinetic interaction discussed in Chapter 1. The upper part of figure 3.29 shows a wavepacket consisting of seven crests and six troughs of negative electrostatic potential. We make the approximation for the moment that this wavepacket travels as a continuous entity in the x direction at a parallel phase velocity which we denote, for this analysis in just one velocity dimension, as v_ϕ. We examine events from the frame of reference of the wavepacket, corresponding to a velocity of zero in the lower part of the figure. Let us see what happens if an electron **1** moving more slowly than the wavepacket is caught by the wavepacket. In the wavepacket frame the electron in question has a negative velocity, say $-\delta v$. All we need to ask now is whether the electron has enough kinetic energy in the wavepacket frame to overcome the potential barrier of the (negative-potential) crests. The test is whether or not

$$\delta v^2 = 2e\Phi_{max}/m.$$

Electron **1** has insufficient kinetic energy. Although it is able to surmount the first two crests, it cannot get past the third. Its velocity in the wave frame becomes reversed, just as for a object colliding with an impenetrable barrier. The outcome seen from the frame in which the wavepacket has velocity v_ϕ is that electron **1** is accelerated from $v_\phi - \delta v$ to $v_\phi + \delta v$. This represents an increase in kinetic energy of

$$\delta E = 2mv_\phi \, \delta v$$

just as for the analogous bat-and-ball effect discussed in Chapter 1 and similarly for a surfer who has successfully 'caught' a wave.

Electron **2** has more than enough kinetic energy. Indeed its velocity in the wave frame is still substantial even at the two negative peaks of the wavepacket. This electron successfully negotiates the hurdles to emerge unchanged in velocity (in any frame) after the wavepacket has passed. This is the equivalent of a surfer who has been left standing by a wave.

Electron **3**, although initially travelling faster than the wave, is unable to overtake it. Reflection in the wave frame causes it to lose kinetic energy, mirroring the process by which electron **1** was accelerated. This is equivalent to our earlier drop shot in tennis, or to a would-be surfer who has not quite understood the objective.

Electron **4** catches up the wavepacket, passes it, and continues on unscathed.

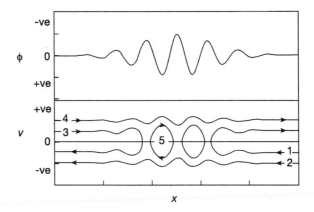

Figure 3.29. Interactions between electrons and wavepackets. Electrostatic potential (upper panel) and electron velocities relative to the wavepacket (lower panel) are shown as functions of position in the frame of reference of the wavepacket. Potential is shown negative uppermost in accordance with negative excursion representing potential barriers. Electrons **1** and **3** have insufficient energy in the wave frame to negotiate the potential barriers, **2** and **4** are able to pass through the wavepacket, while **5** is trapped in a potential well. In this frame, none of the electrons experience a net change in speed. In a frame of reference moving more slowly than the wavepacket, i.e. one with a negative velocity in the lower panel, electron **1** gains kinetic energy from the interaction while **3** loses energy. The energy gain by **1** may be likened to surfing. There is no net change to the energies of **2**, **4** or **5** in any frame (from [160]).

Electron **5** finds itself trapped between the highest pair of potential barriers in the wavepacket with insufficient kinetic energy in the wavepacket frame to surmount them and escape. Electrons could also, in principle, be trapped in the weaker potential troughs if they were travelling close enough to the wavepacket phase velocity when a wave developed around them.

Discussion of the interaction would not be complete without mention of the effect of the above exchanges of energy on the wavepackets. It goes without saying that energy gained by electrons will be lost by the wavepacket, and vice versa. Since the energy of the wavepacket resides in the coordinated oscillatory motion of the plasma, this oscillatory motion will be reduced or augmented by energy gains and losses, respectively, of the electrons. Reduced oscillation will result in smaller potential excursions, i.e. a smaller-amplitude wave, thus reducing, or 'damping down' the wavepacket. Energy losses from electrons would lead to wavepacket growth.

3.4.3.3 Modification of electron velocity distributions

Continuing to refer to figure 3.29, we now consider what effect wavepackets might have on a whole distribution of electron velocities. We'll give all wavepackets the same phase velocity v_ϕ continuing still to make the approximation that they move forward as unchanging entities with this velocity. We shall consider what happens if electron velocity-space densities fall off towards higher velocities. To start with, we'll consider the effect on a one-dimensional electron velocity distribution, which falls off linearly with velocity in the region of interest.

Electrons below a certain velocity (equivalent to a velocity between those of **1** and **2**) will not change their energy, ensuring that the velocity distribution remains unchanged at these low velocities. Similarly, there will no change for velocities higher than those corresponding to a value lying between those of **3** and **4**.

Electrons of type **1**, with initial velocities just below v_ϕ will be accelerated to a new velocity the same amount above v_ϕ. Electrons of type **3** will, by the same token, move in the opposite direction in velocity space. If there are more electrons of type **1** than of type **3**, as there are in the distribution we are examining, the probability is that there will be a net migration towards higher velocities. The migration will proceed until velocity-space densities at corresponding velocities either side of v_ϕ are equal. When this stage is reached, resonant electrons will still continue to gain and lose energy, but the numbers doing so will balance, and there will be no further change in the velocity distribution. If the range of resonance is relatively narrow, so that velocity-space densities fall linearly with velocity in this region, the levelling process will, on this reasoning, create a plateau extending over the full range of resonance. It is very encouraging to note that this is in accordance with Landau's prediction illustrated in the earlier figure 1.17.

Fortified by finding ourselves, as a result of these very basic considerations, still safely on course, we can try to explore a little further. If we were to introduce, in place of a discrete wavepacket velocity, a distribution of wave velocities, the plateau would widen as in figure 3.30. The broad plateau that might, in principle, form by this process naturally brings to mind the broad plateaux seen earlier in some auroral electron distributions.

So far, we have considered the effects of wavepackets of such (small) amplitude that the electron velocity distribution was approximately linear over the velocity range of resonance of a single wavepacket. We can now try to reason what happens at larger amplitudes. Figure 3.31 shows a velocity distribution, approximated for this exercise as a histogram of discrete velocities, subjected to wavepackets of a single phase velocity v_ϕ. The distribution is of power-law form, and so appears as a curve on this linear scale. At low amplitudes the wavepackets transport electrons between the two columns **aa** adjacent to v_ϕ. They level these two columns and create a narrow ledge or plateau in the

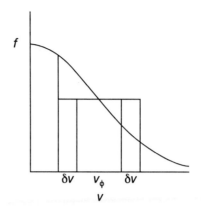

Figure 3.30. Extension of Landau plateau from the $2\delta v$ of figure 1.17 by a finite range of wave phase velocities v_ϕ.

manner already discussed. Larger amplitude wavepackets would be capable of levelling the next pair of columns **bb** as well. Since the velocity distribution in this example is such that the average of the velocity-space densities of these outer columns is approximately the same as the average of the innermost pair of columns, the result of the levelling is simply to extend the plateau. Larger amplitudes still would bring the next pair of columns **cc** into resonance, and bring their velocity-space densities to the same value. However, the average of these is, due to the shape of the original non-linear velocity distribution, noticeably greater than the averages of the inner pairs. The result of the levelling in this case is to produce a plateau with a slight rise at both ends. A still larger amplitude produces more pronounced rises at the ends of the plateau (**dd**). The rise at the upper end can fairly be described as a peak. It appears, therefore, that the process of levelling can, at least in principle, lead to the generation of a peak in the electron velocity distribution—a peak forcibly reminding us of the characteristic peak of the auroral electron distribution. Since it is only in the direction parallel to the magnetic field that lower-hybrid waves match electron velocities, the peak generated by this process would be magnetic-field aligned as it needs to be if it is to correspond to that of auroral electron observations.

Continued exposure to wavepackets of a single velocity would leave the equilibrium, or steady-state distribution and its peak unchanged. A distribution of wavepacket velocities would, though, according to the reasoning above, erode the peak and create ultimately a plateau extending over a velocity range of the wavepackets plus twice the velocity range of resonance.

The finding that a field-aligned peak could in principle be generated by random interactions with lower-hybrid waves forms the basis of the wave theory of auroral electron acceleration. Appendix 3 contains a code to enable you to verify this key conclusion for yourself and to explore the development of velocity

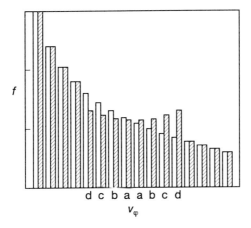

d c b a a b c d

v_φ

Figure 3.31. Generation of a peak in a velocity distribution by random action of waves on an initial monotonic distribution (open histogram). Small-amplitude waves of phase velocity v_ϕ give rise (shaded histogram) to the plateau **aa**. As the wave amplitude is increased, via **bb** and **cc** to **dd**, a significant peak has formed. Entropy increases throughout.

distributions subject to such influences.

3.4.3.4 *Group velocity*

Wavepackets do not, in general, and certainly not in the case of lower-hybrid waves, travel as unchanging entities. They constantly form, disperse and re-form in localised regions as the result of different frequency components travelling at different velocities. This can be illustrated dynamically using figure 3.32. This figure represents two wave trains of slightly different wavelength. If these wavetrains are photocopied onto transparencies and superimposed, beats will be clearly seen.

If the waves so represented were sound waves, the phase velocity would be the same for both frequencies, and propagation of the waves may be simulated by moving both wave-trains sideways at the same velocity. The pattern of beats will, naturally, move at the same velocity, representing wavepackets travelling as continuous entities. The velocity of wavepackets or other travelling modulations is known as the group velocity of the wave.

For lower-hybrid waves the phase velocity increases as wavelength increases and frequency falls. The effect of this can be seen by moving the pattern with the longer wavelength relative to the other. The beats will be seen to move in the opposite direction to the movement. This corresponds to a beat, or group, velocity which is lower than the velocities of either component wave. As a result, lower-hybrid wavepackets travel more slowly than the phase of the

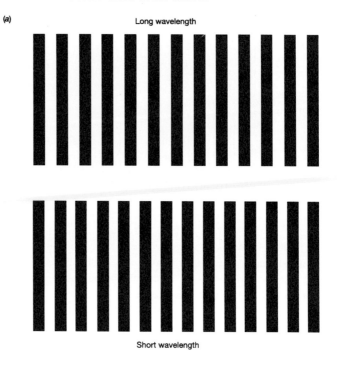

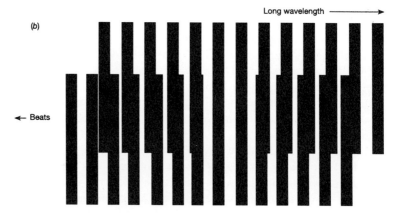

Figure 3.32. Generation and propagation of beats. Trains of crests and troughs valleys of long wavelength and short-wavelength waves are represented by vertical bars. If we place ourselves in the frame of the lower wave and draw across it, as indicated, a transparency of the upper wave, a train of beats (in the form of dark and light patches) will be seen to move in the opposite direction. This demonstrates that, for lower-hybrid waves and others that propagate faster at longer wavelengths, the group velocity is lower than the phase velocity of its components.

oscillations contained within them. Equivalently, oscillations of phase move faster than their envelopes. It is readily demonstrated by these means that for waves which travel faster at shorter wavelength, the group velocity is greater than that of either component.

The parallel group velocity of lower-hybrid waves is related to its parallel counterpart by

$$v_{g\parallel} = v_{\phi\parallel}(1 - \omega_{lh}^2/\omega^2).$$

At the lower-hybrid resonance frequency, the parallel group velocity is zero. At $\sqrt{2}$ times the resonance frequency it rises to a maximum of half the parallel phase velocity. At several times the resonance frequency, it approaches the parallel phase velocity. These characteristics introduce a new class of electron interaction to our previous set of five, and remove one of the original set. Can you see, before reading on, which is removed, and what the new class is?

With the wave-train simulation in mind let us look again at figure 3.29, putting ourselves in the frame of reference of the wavepacket so that the envelope of potential oscillations stays fixed. The pattern of crests and troughs will then move to the right at the difference between the phase velocity, with which it is travelling, and the group velocity, with which we as observers have chosen to travel. Imagine an electron moving to the right and catching up with the wavepacket at a velocity above the group velocity but very close to the phase velocity. If it catches the phase of the wave just right it will be able to travel in a wave trough completely through the wavepacket without at any stage being required to surmount any of the (negative) crests. It would be something like negotiating a revolving door with hands in pockets. If the phasing is unfavourable or the velocity differs too much from the phase velocity, the electron will then run into or be caught up by the potential barriers. The revolving door analogy in this case does not bear thinking about.

The new class of interaction, then, is that of electrons travelling at a velocity between the group velocity and the phase velocity and having their velocities altered by random buffeting from the potential barriers. This new class of interaction is the main type of interaction caused by waves close to the lower-hybrid resonance. Since there are now no permanent troughs of electrostatic potential, our original class **5** of trapped particles finds itself without a home.

For the higher-frequency waves, resonant reflections of types **1** and **3** are more likely. In general there is a mixture of the two, the balance depending on the relative numbers of lower- and higher-frequency waves in the wave velocity distribution.

The essential features of the wave model [161] are shown in the set of figures 3.29, 3.33 and 3.34. Figure 3.33 is the equivalent for the wave theory of figure 3.21 for the potential-difference theory. The model envisages electrons encountering, during precipitation, a region of electrostatic turbulence represented by wavepackets of lower-hybrid waves. The wavepackets vary in numbers, amplitude and velocity. Electrons encounter one or more wavepackets, on their traversal of the region of turbulence. Encounters may or may not be resonant. They may lead in some cases to acceleration (as class **1**), retardation (class **3**) or random change of velocity (class **6**).   Figure 3.34 summarises the types of interaction recognised by the present model of the mechanism, incorporating the properties of lower-hybrid waves as just described. It is the model's engine room.

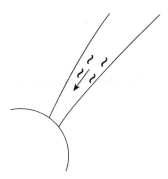

Figure 3.33. Essence of the wave model of the aurora, in which precipitating electrons (arrow) are accelerated by lower-hybrid wave turbulence.

In addition to the interactions with wavepackets, there is a continuous re-adjustment of kinetic energy between parallel and perpendicular motion due to the gradient in the magnetic field. This effect allows some electrons to penetrate to the atmosphere and causes some to mirror, just as it does in the potential-difference model.

The essentially random nature of the process envisaged here, does not, perhaps surprisingly, actually preclude a systematic analysis such as employed for the orderly fields envisaged by the potential-difference theory.   In fact some early progress was made using such a scheme [23].   However, a numerical approach which follows electrons individually through the buffeting by wavepackets and mirroring by magnetic fields is more flexible and is more conducive to the testing of different starting conditions. Models using this type of approach in which on-the-spot decisions are taken by random numbers whose probability of selection is based on the prevailing probabilities are known as Monte Carlo models. Those of Appendices 1–3 are of this type.

3.4.4 Predictions

What would we expect to see from this process in operation?

Plate 4 shows a sequence of contours of electron velocity-space density in $v_\parallel$, $v_\# - v_\perp$ space. There are eight distributions **A–H**, spaced uniformly between altitudes of approximately 24 000 km and 1000 km. The right-hand side of **A** is the source distribution. It is a power law of exponent −3.5 in energy and is

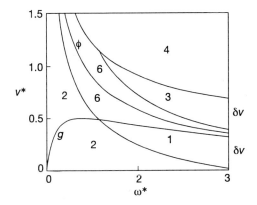

Figure 3.34. Resonance conditions for electron interactions with lower-hybrid wavepackets. The ordinate v^* is velocity, of electrons or waves, measured in units of the wave perpendicular phase velocity, U multiplied by the square root of the ion-to-electron mass ratio. The abscissa, ω^* is (angular) frequency measured in units of the lower-hybrid resonance (angular) frequency, ω_{lh}. Parallel phase (ϕ) and group (g) velocities are shown as functions of ω^*. The amplitude of the wave is indicated by the spacing of the curves on either side of ϕ at the tolerance for electron resonance, δv. Interactions between wavepackets of frequency ω^* and electrons of velocity v^* are identified by the same notation as in figure 3.29. Because of the difference between phase and group velocities 'surfing' by electrons **1** is restricted to transitions between the lower and upper horn-shaped areas. There are now no trapped electrons, but a new class **6** of 'revolving door' electrons has emerged (see [161]).

isotropic over the downward hemisphere.

Wave turbulence is, for this illustration of the model's properties, introduced within the narrow range of altitudes 2000 km immediately below **A**. The turbulence takes the form of wavepackets with angular frequencies mainly in the range 2–5 ω_{lh} where type **1** and type **3** interactions predominate. Wave frequencies continue, though, down to 1.4 ω_{lh} where type **6** interactions are more likely. Wavepacket velocities accordingly range from 17 500 to 80 000 km s^{-1}. Amplitudes are distributed uniformly between zero and 730 V (corresponding to resonance widths of zero to 16 000 km s^{-1}). The numbers of wavepackets are such that 3 keV electrons, for example, would encounter on average just 2.6 wavepackets, resonant and non-resonant, on traversing the region of turbulence.

The right-hand side of **A** shows the isotropic source distribution. At **B**, just below the turbulence, the downward half of the velocity distribution is considerably modified with respect to the source. A well marked field-aligned peak has appeared at $v_\parallel = 34\,000$ km (3 keV), and the yellow contour has extended along the $v_\parallel$ direction to engulf it. This is caused by type **1** acceleration, especially those effected by the largest waves.

By **C** there has been little further change, as expected since there has been no further turbulence, and the magnetic field strength has increased by less than 50%. We can now see quite clearly the broadening of the electron distribution, i.e. the heating that has resulted from type **6** interactions in the region of turbulence.

Continuing down through the other altitudes to **G**, the field-aligned peak is seen to extend gradually along a circular path (the cross section of a spherical shell) almost as far as pitch angles of 90°. This is caused by magnetic mirroring in the field which has by this altitude increased to 16 times its value at **B**.

By **H**, the downward part of the distribution is completely isotropic. Electrons which were at the fringes of the peak at **B** have had their pitch angles increased to 90° and beyond, producing a downward-hemisphere isotropic shell of velocity-space densities at 3 keV, and causing some of the electrons to mirror and set off in an upwards direction. Before we follow them, we just note that the central, low-velocity region of the electron distribution is unchanged from that of the original source distribution. This is because, with the wavepacket velocities and amplitudes used in this example, electrons below 1750 km s^{-1} (9 eV) do not resonate—they do not catch the waves.

There are two populations of electrons to follow upwards. We'll take the mirroring, higher-energy ones first.

At **G**, the electrons mirroring below have had their pitch angles reduced by the diminishing magnetic field so that they remain at the edge of the (upward) loss cone. The upward-flowing electrons at pitch angles close to 90° are those that mirrored immediately below **G**. These are from the still wider fringes of the initial beam, and so have lower velocity-space densities than those closer to the loss cone which originated nearer to the centre of the beam.

As we travel on upwards to **A**, we see that the high velocity-space densities at the edge of the loss cone lie on the surface of a cone which gradually closes, like the loss cone itself, due to the weakening field at higher altitude. The feature created in the electron velocity distribution could well be described as a conic. The wave model produces, therefore, a conic which is simply the mirrored fringe of the beam. Apart from the appearance of a loss cone, the distribution at low velocities remains unchanged and isotropic.

We return to **H**, and to the electrons flowing upward at pitch angles between 120 and 180°, i.e. inside the loss cone. These electrons result from an introduction into the model of a backscattered and secondary albedo from the atmosphere. The actual distribution introduced could in principle be computed, along the lines of the code developed by Evans, from the distribution that the model injects into the atmosphere. However, since this would be a major task and would add very little to our understanding, we have introduced a nominal albedo approximating that observed experimentally, in order to complete the picture. As a result, relative velocity-space densities inside and outside the loss cone have no meaning in this version of the model. It is meaningful though for the loss-cone electrons to occupy a diminishing cone of directions as greater altitudes are reached.

The advisability of looking at measurements or the results of computations in more than one framework is emphasised with the inclusion of plate 5, which gives the corresponding energy–pitch-angle spectrograms of expected detector count rate. A new feature, seen most clearly at **A**, is a depletion region (blue) for low energies (< 1.1 keV) at small pitch angles. This is simply a manifestation of the depletion seen at the lowest energies of resonance in the upper panel resulting from the net migration from here into the peak which stands out very sharply at higher energies. This field-aligned depression is the wave theory's equivalent of the 'hollow' sometimes noticed in this region of the electron distribution, and illustrated earlier in figure 3.18.

At lower altitudes the depletion region widens in pitch angle, in line with all other features. In common with the other features, too, the depletion region is reflected to create an extended loss cone at low energies which is much wider than the nominal loss cone. This commonly observed feature is, then, another to be expected from wave turbulence.

It is worth noting too that the heating caused in the wave model by wavepackets with frequencies close to the lower-hybrid resonance shows up very clearly in the extension of count rates over a broadening range of energies.

3.4.4.1 *Versatility*

The versatility of lower-hybrid waves in being able to generate a wide range of different features, depending on the composition of the waves and their location, can be appreciated quite readily from plates 4 and 5.

If the wavepackets were located at a lower altitude than just below **A**, the field-aligned beam would penetrate down to lower altitudes, possibly as far as the atmosphere. A similar effect would result from spreading the wavepackets out over a wider range of lower altitudes. Another consequence would be that fringes of the beam would be less likely to mirror, so weakening the conic.

If the wavepackets were of lower frequencies than those used above, the peak would be weaker and the heating would increasingly become the main result of the turbulence. If located at high altitudes lower-frequency waves would produce a broad plateau over a wide angular range. At low altitudes they would draw out electron velocities along the $v_\parallel$ axis, in the manner seen earlier in figure 3.10.

It will have been clear from figure 3.31, and can be verified by experiments carried out using the code in Appendix 3, that a few large-amplitude wavepackets are much more effective in altering the electron velocity distribution than are any number of small-amplitude ones.

It will also be clear that the region of turbulence needs to contain enough wavepackets for a significant fraction of the traversing electrons to be modified. If the numbers are too great, though, key features such as the peak will be eroded by increasing entropy.

If a proportion of the wavepackets had upward parallel velocities, electrons would be accelerated in the upward direction also. A symmetrical distribution between upward and downward velocities would, naturally, lead to counter-streaming, another of the phenomena we noted earlier. The preference for upward directions which we have seen in some examples would, in this theory, correspond to a bias favouring upward parallel velocities. The predominant downward acceleration occurring in auroral electron streams argues though in favour of the normal preference being downward. While this observed bias can certainly be accommodated in the theory, it does not follow as a natural consequence. This is an important issue waiting to be resolved in further developments. Swift concluded that the increase of number density towards lower altitudes would create a preference for downward propagation. The increasing magnetic field might also play a role. The powerful effect that gradients in a medium might have on wave motion is brought to mind quite forcibly when we note, for example, how the shelving of a beach produces inward, never outward, breakers.

Another key issue is that of multiple peaks, such as those we saw in figure 3.8. While it is perfectly possible, in principle, for there to be a region of turbulence at high altitude leading to an isotropic peak in downgoing electrons just above the atmosphere, as in the example illustrated in plates 4 and 5, and for there to be at the same time a region of lower velocity wavepackets at low altitude contributing independently a field-aligned peak at lower energy, such a double accelerator does not arise directly from the theory.

3.4.5   Wave fields

The study we have just carried out demonstrates that lower-hybrid wavepackets would, if they were large enough in amplitude and present in adequate numbers, produce most, possibly all, of the features observed in auroral electron acceleration. The question that now remains is—do the waves exist? A prerequisite is to know what kinds of amplitude we are looking for.

3.4.5.1   Estimates of field strength

The strongest electric field found in the wavepackets employed in the above analysis can be estimated straightforwardly. At a nominal frequency of 3 kHz and a representative parallel phase velocity of 30 000 km s^{-1} the parallel wavelength would be 10 km, making the perpendicular wavelength approximately 250 m. The electric field would peak twice per cycle of 0.33 ms, with the highest positive and negative values being

$$\mathcal{E}_{max} = 2\pi\,\Phi/\lambda_\perp.$$

With $\Phi = 730$ V, $\mathcal{E}_{max}$ would be 18 V m^{-1}.

Such fields would be detected as fleeting (< 0.1 ms) positive and negative excursions on instruments able to take and transmit readings at 10 000 or more per second.

Instruments with lower resolution would still be able to detect wavepackets of the above description through the average, or rather the square root of the mean square, rms, field strength in a wavepacket as a whole. While the simple average taken over the full set of oscillations will be zero, rms provides a useful measure of the overall magnitude of the field. The rms electric field for a wavepacket as whole for the above wavepacket would be approximately 10 V m^{-1}. Such fields would be detected for periods of a few wave cycles, i.e. for periods lasting around 1 or 2 ms. Instruments able to measure the rms field at the rate of a few thousand samples a second would be needed to detect them. This is still quite a challenging demand, and has not yet been met in the acceleration region.

To estimate how the wavepackets would look to instruments with lower resolution we have to allow for the intervals between wavepackets where the field is zero. To calculate this we need to know how closely the wavepackets are packed. The greater the vertical extent of the turbulent region the more loosely the wavepackets need to be packed to give the same overall exposure to particles travelling through. These key factors can be taken into account to arrive at an empirical relation for the overall rms electric field. In a turbulent region 10 000 km in height permeated by wavepackets containing six oscillations, as in figure 3.29 the proportion of the total volume containing wavepackets, would, at any one time, be less than 1%, making

$$\mathcal{E}_{rms} = \sqrt{E_{peak}}/500 \text{ V m}^{-1}$$

where E_{peak} is the energy where the peak occurs. So, for a peak at 10 keV, $\mathcal{E}_{rms}$ is estimated to be 200 mV m^{-1}. The estimate would double if the acceleration region was only a quarter of the 10 000 km adopted for this estimate, and halve if it were four times as great. It would be doubled if wavepackets had 20 oscillations, and halved if there were only one. The estimate would double if the plasma density were 16 times higher than the 10^5 m^{-3}, assumed here, and halved if it were 16 times lower. This would be the rms field detected continuously in low-resolution measurements, however long the period over which they were averaged (providing that the detector remained in the turbulent region throughout the measurement, of course).

Since the wave crests of lower hybrid waves are aligned approximately perpendicular to the magnetic field, the 200 mV m^{-1} deduced for a substantial peak at 10 keV applies to this perpendicular direction. The rms electric field parallel to the magnetic field will be 42 times smaller, i.e. 5 mV m^{-1}. A similar peak at 3 keV instead of 10 keV would, according to this estimate require an $\mathcal{E}_{rms}$ of about 100 mV m^{-1} in the perpendicular direction, for which the corresponding parallel field is 2.5 mV m^{-1}. Naturally, the field strengths required to produce less pronounced peaks would be correspondingly less than these estimates.

3.4.5.2   *Significance of estimated fields*

We will now consider the implications of these electric fields.

The energy density, or energy stored in or required to establish an electric field in a cubic metre of space, is

$$\Xi = \tfrac{1}{2}\varepsilon_0 \mathcal{E}_{rms}^2$$
$$= 4.43 \times 10^{-12} \; \mathcal{E}_{rms}^2 \; \text{J m}^{-3}.$$

The above estimate for a substantial peak at 10 keV leads to, for the waves

$$\Xi_{waves} = 2.0 \times 10^{-13} \; \text{J m}^{-3}.$$

Since local, or ambient particle densities in the acceleration region are, say, 3×10^5 m^{-3}, and mean energies are around 3 keV (taking into account both electrons and ions), the energy density of the plasma particles is, typically,

$$\Xi_p = 1.5 \times 10^{-10} \; \text{J m}^{-3}.$$

The estimated wave energy density is then much less than this, in fact approximately one-thousandth of it. However, such a value, although seemingly small, is not negligible—far from it. In fact, plasmas with wave fields of such relative energy density are understood to be in a state in which particle velocity distributions are strongly influenced by the electric fields—a state known as strong turbulence [162]. Wave turbulence is considered weak only when the wave energy density is perhaps less than one-millionth of the particle energy density. In the practical example of the heating of electrons in tokamaks, discussed earlier, the wave energy density is usually less than one-hundred-thousandth of the plasma energy density. Since we are suggesting in the wave theory of auroral electron acceleration that electron distributions are strongly modified by the waves, it is very reassuring to find that the estimated fields are strong enough to constitute strong turbulence.

The quarry is now quite well defined. We are looking for lower-hybrid waves, whose frequency in the acceleration region is of the order of a few kilohertz, with peak magnitudes at times in excess of 10 V m^{-1} and with rms amplitudes somewhat less than 1 V m^{-1}.

3.4.5.3   *Wave measurements*

When we first realised the possible importance of lower hybrid waves, we asked at international conferences and elsewhere whether waves of this type had been detected. The answer was no, they had not. This seemed at first something of a show stopper. However when we pursued the question by asking whether the experiments that had been performed were capable of detecting and identifying them, the answer was no, they were not. They had not been seen, it was true, but then no-one had looked for them!

There is still nothing approaching a complete survey of either real space or frequency space to give a proper indication of the true situation. However, there is every reason to believe that one of the earliest measurements of turbulence in the auroral acceleration region (by Scarf and his colleagues [163]) was in fact one of lower-hybrid turbulence. Frequencies lay in the range between one and a few kHz. A sharp cut-off at the lower frequency suggested very strongly that this was the lower-hybrid resonance frequency. More recently there have been many reports of strong (several hundreds of mV m^{-1}) turbulence identified as lower-hybrid turbulence [135, 164]. Wavepackets, observed in terms of electric field oscillations, but which correspond very closely to the oscillations in potential in our hypothetical wavepacket of figure 3.29, have recently been detected as figure 3.35 shows. However, these are not themselves candidates for auroral particle acceleration because the amplitudes are far too low and they were, in any case, seen below the acceleration region. It is indicative of the difficulty of the problem that measurements, as here, are perforce made in terms of a local quantity—electric field, while interpretation requires an extended property— potential difference.

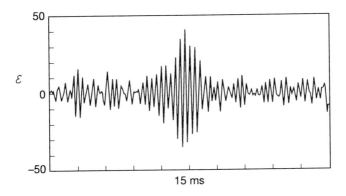

Figure 3.35. Electric field (in mV m^{-1}) of lower-hybrid wavepackets seen below the auroral acceleration region by the Freja satellite (from [165]).

A real breakthrough came during the final stages of preparation of this manuscript with the release of high-resolution wave measurements made by the FAST satellite within the acceleration region. These measurements indicate that waves answering the description of lower-hybrid waves—primarily electrostatic waves in a broad band at and above the lower-hybrid resonance frequency—are found in association with downward-accelerated auroral electrons [166, 167]. In the former case the wave energy density can be estimated from the published results to be as high as 10^{-13} J m^{-3}, which, together with the peak in the electron distribution in the neighbourhood of 10 keV, corresponds exactly with the earlier-mentioned requirements for the wave model. In the latter case there is a somewhat lower wave energy density and a lower peak energy, again as anticipated. The stop press news is, then, that the model's requirements for wave energy density now appear to have been met in practice.

3.4.6   The 'chicken-and-egg' question

A question that arises with wave–particle interaction is—did the waves accelerate the particles, or did the accelerated particles produce the waves? The wave theory depends upon the former, of course, but the latter may also be operative. It has been suggested in fact that an electron beam, such as the archetypical auroral one, will generate lower-hybrid waves which will be absorbed by perpendicularly moving ions to create auroral ion conics [168, 169]. How can we tell, then, even if it can be established that there is a causal relationship, which way energy flows? If we try to adopt the stance that the waves are produced by the electron beam, we are then left in the unsatisfactory position of leaving the generation of the beam (ruling out conservative processes) to an as yet unspecified mechanism. If, however, we can identify another source for the waves, as we shall attempt to do below, it may be possible to accommodate both senses of energy flow into the same model, seeing flow from waves to electrons as the primary process and any reverse flow as a form of feedback serving to keep the peak within bounds.

3.4.7   Power source and energy budget

It is important, if we are to appreciate the process fully, to know what gives rise to the waves and where the energy comes from.

Auroral electrons carry energy into the upper atmosphere at a rate which can be estimated from the energy distributions to exceed at times 100 mW m^{-2}. If the waves are responsible for this, in a continuous or steady-state process, they must deliver in the acceleration region 100 mW for each square metre of the acceleration region at atmospheric heights. They must also be replenished at this same average rate.

We can see very quickly that replenishment cannot be a long, slow process working at relatively low power. The volume of a truncated-pyramid-shaped element of the acceleration region having an area of 1 m^2 at the atmosphere and being 10 000 km in height (see figure 3.36), is $\sim 5 \times 10^7$ m^3. The estimates of wave energy density needed to effect the required degree of acceleration ($\sim 2 \times 10^{-13}$ J m^{-3}) combines with this to give a total, wave energy in a 1 m^2 element of the accelerator as $\sim 10^{-5}$ J. This energy, used at the rate of 100 mW, would last for just 10^{-4} s. This is of the same order as the period we estimated for the lower-hybrid waves; in fact it is rather less for those close to the lower-hybrid resonance frequency. The clear indication is that, if the acceleration really is driven by waves, the waves will have lifetimes so short that they will complete perhaps only a partial oscillation before being attenuated—Landau damped—by the electrons.

The fact that the wave energy was too low to power auroral electron acceleration for more than a small fraction of a second was, and still is in some quarters taken to eliminate the proposition altogether. However this reasoning does not allow for the possibility of replenishment. After all, it is not the size of a reservoir that determines the rate at which a steady supply can be taken from it, but the rate at which it can be re-filled.

3.4.7.1 Replenishment

The question that has to be asked, not just for the wave theory but for any theory, is—can the energy reservoir be replenished at the rate at which energy is used in a strong aurora, i.e. at ~ 100 mW for every projected square metre?

We saw at the beginning of this chapter that a typical north–south thickness of the stream of electrons producing an auroral 'inverted V' structure is around 100 km. That is, the north-south extent through which power flows in the form of accelerated electrons is around 100 km. The degree of acceleration is not uniform over the structure, as we saw, and the average power carried by the electrons in a strong aurora might be, say, 50 mW m^{-2}, rather than 100 mW m^{-2} over the full width. Consider the truncated inverted pyramid of figure 3.36, a nominal 10 000 km in height, with a (horizontal) surface at the base 100 km north–south by 100 km east–west, i.e. a base area of 10^{10} m^2. The power flowing out through this surface would, using the above figures, be

$$P = 500 \text{ MW.}$$

To maintain the steady state we know the aurora can exhibit for at least tens of minutes, this amount of power must enter through one or all of the remaining surfaces. It cannot have been through the east or west faces, since this would serve only to rob the neighbouring regions which have also to be powered. This leaves the top surface and the north and south vertical surfaces. The area at the top of the inverted pyramid is, in view of the expanding cross section of the tube of force, approximately eight times that at the bottom. If the above power were to enter from the top, then, the power density would be approximately

$$\mathcal{P}_{\text{top}} = 6 \text{ mW m}^{-2}.$$

The north and south faces of the region each have areas 170 times that of the base. If the power had entered the region via one of these, the power density would need to be just

$$\mathcal{P}_{\text{face}} = 0.3 \text{ mW m}^2.$$

Is there a power source capable of meeting either of these requirements?

3.4.7.2 Power source

The only form by which power could enter, if not in the form of waves which ⌐xperiments seem to eliminate anyway, seems to be as particle kinetic energy.

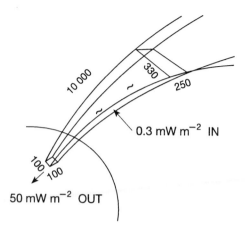

Figure 3.36. Energy budget for the wave model. A stylised volume element of the acceleration region has a cross section at the base of 100×100 km, and, because of the geometry of dipolar magnetic field lines, a cross section of 330×250 km at 10 000 km. Power is carried in by plasma sheet ions at 0.3 mW m^{-2} over one of the near-vertical sides, and carried out by accelerated electrons at the (170 times smaller) base at 50 mW m^{-2}.

Particle number densities (10^5–10^6 m^{-3}) and energies (~ 1 keV) in the plasma sheet are such that the typical power density in the ions and electrons impinging on any unit area is ~ 1 mW m^{-2}. While this power density flowing through the top surface would be too low to replenish the region, it would be perfectly adequate if applied over either or both of the north or south faces.

There is, in fact, very good reason to suspect that these faces might be important, even instrumental in creating the discrete aurora. One of the many striking features of a bright auroral arc, noted earlier with reference to the cover's central panel, is its frequent location between a glow aurora on the equatorward side and a much darker sky towards the pole [89, 170, 171]. The fact that this configuration is not unusual seems to indicate that the acceleration responsible for the arc is a boundary phenomenon. Latitudinal traversals of the particle streams by rockets and satellites certainly reinforce this conclusion, to which plate 2(*a*) bears witness. We have long suggested [172] that the boundary in question is, in fact, a boundary layer, possibly the plasma-sheet boundary layer (PSBL) between the plasma sheet and the lobes of the Earth's magnetotail. Across the PSBL there are, naturally, gradients in particle density and energy and in magnetic field strength. The gradients are steeper closer to the Earth where the magnetic lines of force defining them are closer together. The last point is in accord with the near-Earth, rather than deep magnetosphere, location of the acceleration region. Natural systems abhor gradients or any form of spatial structure, just as they object to structure in velocity space. These gradients might be expected therefore to be unstable and give rise to processes or 'instabilities' which in promoting entropy would generate waves, possibly lower-hybrid waves.

3.5 Summary

The above discussion has brought us to the front line in the development of the wave theory of auroral electron acceleration. In summary, it is that the juxtaposition of the different plasmas of the plasma sheet and the tail lobes, via the intervening boundary layer of the PSBL, causes energy to be drawn from plasma-sheet ions and electrons making excursions into the boundary layer. The non-uniformities in velocity distribution and magnetic field give rise to lower-hybrid waves which accelerate electrons traversing the region of turbulence. It is interesting, in view of our earlier discussion in Chapter 1 of practical particle accelerators in the laboratory, to note that the same multi-stage process is put to practical use in tokamaks heated by neutral ion injection [150]. Here, neutral particles are fired into a plasma, becoming ionised as they enter. The ions moving perpendicular to the magnetic field give rise to lower hybrid waves, which then accelerate electrons parallel to the magnetic field. In this practical application the accelerated electrons carry an electric current whose associated magnetic field helps to keep the plasma particles trapped or 'contained', and their energy eventually finds its way into the plasma's ions.

3.6 A loose end

Before concluding this chapter, we have a loose end to attempt to tie up. This is in connection with the earlier discussed observation by Alfvén in 1958 that electrons became accelerated in discharge tubes containing double layers. In what appears to be a similar experiment, Takeda and Yamagiwa have observed [173] x-rays with energies much greater than that corresponding to the applied potential difference. They also observed considerable wave turbulence. It would be fully consistent with the reasoning above to attribute the electron acceleration responsible for the x-rays to the observed wave turbulence. Such an interpretation would also reinforce the emphasis Alfvén placed later (in 1987) on the importance of waves in double layers [174].

3.7 Conclusions

The wave theory is in principle capable of accounting for most—possibly all—of the properties of accelerated auroral electrons. Recent high-resolution measurements appear to have discovered the envisaged lower-hybrid waves.

The potential-difference theory envisages energy transfer from conservative fields and is, therefore, untenable.

Chapter 4

Electron acceleration at the Earth's bow shock

This chapter is the first of a set of three devoted to the acceleration of electrons of the solar wind caused by obstacles in the path of the solar wind. The obstacle in this first case is the terrestrial magnetosphere, and the acceleration site is the Earth's bow shock.

4.1 The archetypal electron transition

The key facts of the transition from the solar wind to the magnetosheath undergone by the electron energy distribution at the bow shock are summarised in figure 4.1, a much quoted measurement made by the satellite Vela 4 on 5 June 1967 [175]. Electrons in the magnetosheath clearly have a broader energy distribution than their forebears in the solar wind. The change this represents has been found to be generally greater on those occasions and at positions where the interplanetary magnetic field was nearly parallel to the shock surface, a condition known, somewhat confusingly, as a perpendicular or quasi-perpendicular shock.

The archetypal magnetosheath distribution is not only broader than the solar-wind distribution, but is here actually almost flat-topped, perhaps rising slightly towards higher energies. An assumption often made, e.g. [176], and which has implications for the interpretation, is that the magnetosheath distribution remains flat at energies below the limit of measurement (16 eV in figure 4.1) right down to zero energy. In truth, though, we do not know what happens below the lowest energy of measurement, a limit that has been imposed upon all measurements to date somewhere between 10 and 20 eV by the sampling strategy, by contamination from the cloud of photoelectrons surrounding the spacecraft, or from detector response to solar ultra-violet radiation. In an example obtained with the OGO-5 spacecraft very similar to that of figure 4.1, the velocity-space density at the lowest recorded energy, 12 eV, was actually

164

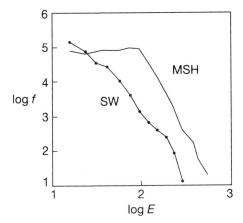

Figure 4.1. Comparison of solar-wind **SW** and magnetosheath **MSH** electron energy distributions. The magnetosheath electrons are more energetic and their distribution has an almost flat plateau from 20 to 100 eV (from [175]).

some three times higher than that of the plateau, but this rise was interpreted as contamination from photoelectrons [44]. In some of the measurements to be examined below there is no plateau, and velocity-space densities will be seen to fall smoothly and monotonically from the lowest energy measured, which was 10 eV.

A spacecraft particularly well equipped to investigate electron acceleration at the bow shock was the AMPTE UKS [39], one of three Active Magnetospheric Particle Explorers [177, 178] taking part in a unique mission, to which the next chapter will testify. The UKS was in a highly eccentric orbit only slightly inclined to the ecliptic plane and with an apogee of 18.5 R_E. It carried a novel type of electrostatic analyser as an energy filter for electrons in some detectors and ions in others for an array of channel plates [179, 180]. The novel feature of the electrostatic analyser was the large angle (270°) through which particle trajectories were deflected between the aperture and the sensor. This large angle gave the detector a high resolution in energy and, particularly welcome for the present application, almost complete immunity from photo-electron contamination. The UKS carried also a magnetometer [181] and wave-measuring equipment [182]. The UKS was able to operate with all experiments running for only four hours at a time because it was too small to carry enough solar cells to power it continuously. The re-charging time was ten hours. Periods of operation had, therefore, to be selected with great care and a degree of good fortune. A telling advantage, though, that this spacecraft had over its predecessors was that its rate of measurement was very high, so that the measurements were made with unusually high resolution and that real-time displays of the measurements were on hand and closely monitored at the control centre of the Rutherford Appleton Laboratory.

By way of orientation, plate 6(*a*) shows in spectrogram form a passage from the solar wind through the magnetosheath to the magnetosphere made in a five-hour period on 4 November 1984. After the transition from the solar wind to the magnetosheath at the bow shock, the electron energy spectrum becomes broader as a result of acceleration and heating at the shock. The location is shown (framed) in figure 4.2 along with all others encountered in the spacecraft's seven-month mission. The average location of the bow shock during this period of relatively high solar-wind speed was somewhat closer to the Earth than usual. The bow-shaped locus of the points is very apparent, stretching out to at least as far as the spacecraft could explore, 18.5 R_E, on the dawn flank. The mean position of the 'nose' is some 13 R_E upstream from the Earth's centre in this epoch. A similar pattern is found by other spacecraft too, with the mean position varying according to solar-wind conditions. Transitions similar to the second one in plate 6(*a*)—from the magnetosheath to the magnetosphere, across the magnetopause—will be the subject of Chapter 7.

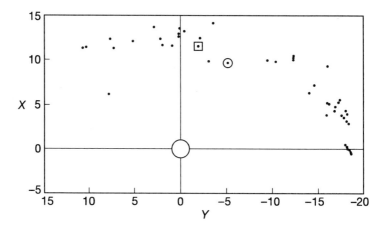

Figure 4.2. Location of bow-shock crossings, all close to the ecliptic plane, made by UKS from September to December 1984. The Earth is indicated by the circle at the origin of coordinates, X is the distance in R_E towards the Sun and Y is the distance in R_E radially outwards on the dusk flank. In this period the bow shock was typically at a distance of some 13 R_E upstream and some 18 R_E on the dawn flank. The crossing depicted in plate 6(*a*) is framed. The case study crossing is ringed (compilation by P A Smith).

4.2   A case study

Plate 6(*b*) is an energy–time spectrogram of electron intensities centred on a particularly clear-cut outbound crossing of the Earth's bow shock on 6 October 1984. The location of this crossing is ringed in figure 4.2. Within just over one minute electron intensities at all energies dropped, as did the magnetic field strength, from magnetosheath to solar-wind values. The spectrogram shows that the drop was preceded by an intensity enhancement lasting for about one minute. The enhancement occurred close to the bottom of the magnetic 'ramp' from 60 nT in the magnetosheath down to 10 nT in the solar wind.

4.2.1   Orientation

Figure 4.3 gives the orientation of several key quantities. At the time of the crossing the spacecraft was travelling directly upstream. Solar-wind ions flowing as expected directly away from the Sun in the upstream region were deflected as they crossed the shock. The surface of the shock was deduced from the flow and field directions to be inclined as shown as the spacecraft crossed it (or as it crossed the spacecraft). The solar-wind magnetic field is inclined at approximately 24° to the shock surface, making this a quasi-perpendicular shock. The magnetosheath magnetic field is only 8° from the shock surface.

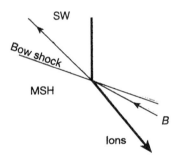

Figure 4.3. Magnetic field and ion flow in the solar wind and magnetosheath close to the bow shock for the case study (magnetic field and shock orientation, courtesy Imperial College; ion flow, courtesy A D Johnstone).

4.2.2   Electrons

4.2.2.1   Inbound

Figure 4.4 shows the energy distributions of electrons travelling in a downstream sense, approximately anti-parallel to the magnetic field. One distribution is the incident solar-wind distribution, one is that near the foot of the magnetic ramp at the point where the magnetic field strength had risen from 10 to 20 nT, and the third is the magnetosheath distribution where the field was 60 nT. A point of correspondence will be seen at once with figure 4.1. The magnetosheath has a broader distribution than the solar wind, but exhibits a sloping shelf rather than a plateau. The ramp distribution is similar to that in the solar wind, except that it displays higher velocity-space densities at all energies, or, equivalently higher energies for given velocity-space densities. The ramp distribution lies above the magnetosheath distribution at high and low energies, but below it in the middle range. The ramp distribution corresponds to the tower-like feature in the plate 6(*b*) spectrogram.

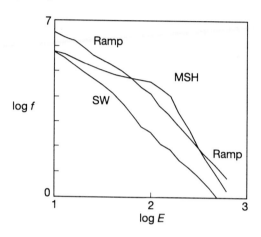

Figure 4.4. Energy distributions for electrons inbound to the magnetosheath, i.e. those guided in the opposite direction to the magnetic field (see figure 4.3) from the solar wind via the 'ramp' into the magnetosheath.

Inbound ramp distributions exceeding those of both the magnetosheath and solar wind at low energies have also been reported by Feldman and colleagues [183]. In a crossing of the bow shock made by the satellite ISEE-2 on 13 December 1977, well representative of many made by this spacecraft, the ingoing ramp distribution showed a slight peak at about 20 eV, rather than continuing to rise towards low energies as in figure 4.4. The magnetosheath distribution exhibited a plateau very similar to that of figure 4.1, and exceeded the ramp and solar-wind distributions from 50 eV to the highest measured energy of 400 eV.

4.2.2.2 *Within the ramp*

Electron distributions within the ramp are shown in figure 4.5. The figure shows the inbound electrons (pitch angles close to 180°) which also appeared in figure 4.4, outbound electrons (pitch angles close to 0°) and those gyrating approximately perpendicular to the magnetic field (pitch angles close to 90°). The last, while neither inbound nor outbound, are destined, because of the converging magnetic field, to become outbound. The distributions are approximately the same, constituting approximate isotropy. Note, however, that there are differences amounting to a factor of two in places. The inbound electrons exceed the others at low energy, while the perpendicular ones dominate at high energy.

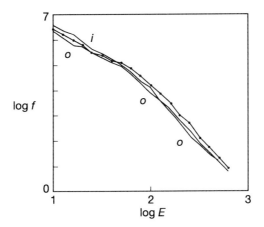

Figure 4.5. Energy distributions for electrons in the ramp: inbound (i), outbound (o) and at pitch angles close to 90° (line with points).

Typical measurements from the ISEE spacecraft [183] show a much greater (almost order of magnitude) excess of inbound electrons over outbound electrons at low energy, and, in contrast to figure 4.5, a clear (factor of two to ten) dominance of outbound over inbound at high energy.

4.2.2.3 *Within the magnetosheath*

Energy distributions within the magnetosheath are shown in figure 4.6. Again, there is approximate isotropy, which is indeed the usual finding. Note, however, that there are differences amounting in places to almost an order of magnitude. The outbound distribution is the highest at low energies, but is the lowest at high energies. The general form is a sloping shelf from 10 to 100 eV, with a steeper fall beyond. The form is almost a lower energy version of the auroral rays of figure 3.5.

The ISEE typical measurements also show approximate isotropy in the magnetosheath. There, however, the distribution forms a plateau from 12 to 150 eV and, in contrast to the above finding, the outbound distribution exceeds the inbound where it falls steeply at high energies.

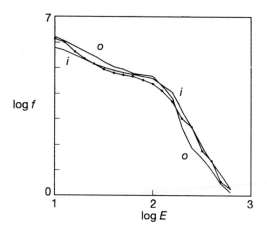

Figure 4.6. Energy distributions for electrons in the magnetosheath; symbols as in 4.5.

4.2.2.4   In the solar wind

Here (figure 4.7) the 'perpendicular' electrons exceed the others at low energies, while the outbound electrons have the highest velocity-space densities at high energies.

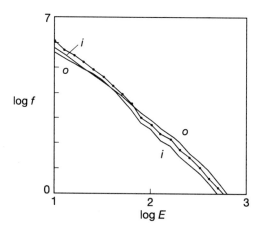

Figure 4.7. Energy distributions for electrons in the solar wind; symbols as in 4.5.

The ISSE-2 results show the outbound electrons to exceed the inbound ones by a factor of three or more over most of the measured range, the velocity-space densities becoming comparable only at and below 25 eV.

4.2.2.5 Summary

The survey of the case study event is completed, and conveniently summarised, by the velocity-space distributions for solar wind and magnetosheath electrons appearing in figures 4.8(*a*) and (*b*). Cross sections of these distributions, taken along $v_\parallel$, $v_\#$ and $v_\perp$ have already appeared in figures 4.6 and 4.7. Figure 4.8 gives the complete picture, albeit with less resolution. The approximate isotropy of both distributions is confirmed by the near-circular contours, as are the biases at high energies in the magnetosheath distribution towards inbound electrons and in the solar wind towards outbound electrons.

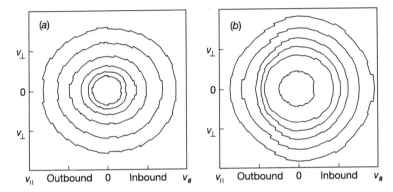

Figure 4.8. Contours of electron velocity-space density in the solar wind (*a*) and magnetosheath (*b*) for the bow-shock case study. Parallel, antiparallel and perpendicular velocities range from 0 to 1.6×10^7 m s^{-1} (corresponding to an electron energy of approximately 16 keV). Contours of velocity-space density range in order-of-magnitude steps from 1 km^{-6} s^3 (outermost) to 10^5 km^{-6} s^3 (innermost) (analysis D R Lepine).

4.2.3 Taking stock

Before attempting to interpret these findings, we shall take stock of the situation to see what type of regime we are working in.

Figure 4.9 takes a wider look at the geometry. It shows schematically, and not to scale, that the magnetic lines of force threading the crossing point of the shock layer within which the changes occur must intercept the layer again somewhere else as well. There are, then, both upstream and downstream crossing points. The solar-wind portions of the tube of force depicted are continuations of one another, since the plasma which they thread has not yet come into contact with the shock. The magnetosheath portion, which links the crossing points, is curved in the sense shown because of the reduction in drift speed effected at the shock. The phenomenon is analogous to that of the refraction of a light ray on entering a medium of increased refractive index, the lines of force behaving geometrically similarly to the wave fronts (as opposed to the rays) of light.

Consequences of the 'refraction' are a narrowing of the tube of force as it penetrates into the magnetosheath and a corresponding increase in field strength (see also figure 2.4).

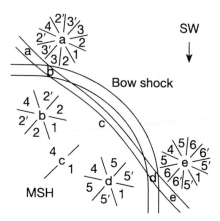

Figure 4.9. Schematic of lines of force and electron populations of the solar wind (**a** and **e**), the ramp (**b** and **d**) and the magnetosheath (**c**). The routes taken by the electron groups **1** to **6** may be followed in the indicators aligned with the magnetic field in the corresponding region. Primes indicate magnetic mirroring. Group **1** enters from **a**, occupies a larger range of pitch angles in the stronger field at **b**, and fills the hemisphere of forward directions at **c** in the still stronger field of the magnetosheath. The sequence is reversed on the outward journey via **d** to **e**. The largest pitch angle members of **2** mirror at **b**, and the smallest are mirrored before reaching **c**. The mirrored electrons from this population become the outbound **2′** electrons at **a** and **b**, and similarly for **3** which mirror before reaching **b**, and for the complementary groups **4**, **5** and **6** approaching from **e**.

To serve as a control for the ensuing discussion of what acceleration processes might be at work let us investigate what would happen if there were no acceleration. In order to do this we shall consider the electrons to travel adiabatically within the system. At this stage we shall assume that there are no electric fields of any consequence.

Consider first, electrons of the solar wind gyrating about and being guided by the undisturbed solar-wind magnetic field towards the upstream crossing point, i.e. electron groups **1**, **2** and **3** in the insets of figure 4.9. Electrons of group **1** are travelling most nearly (anti-) parallel to the field (at pitch angles centred on 180°), those of **2** have medium pitch angles, and those of **3** have pitch angles close to and just above 90°.

On the ramp, where the field has increased above solar-wind values (note the converging lines of force), group **1** extends over a wider range of small pitch angles than in the solar wind. Group **2** will, again because of the increase in field strength extend to near 90°. Electrons of group **3** will not penetrate as far as the ramp but will be mirrored by the field to become **3′** in the solar wind, and so form part of the outbound distribution there.

On moving into the magnetosheath where the field strength is still higher, group **1** now take up the full inbound range of pitch angles (0–90°), while **2** are mirrored to **2′** in the outgoing ramp and solar-wind distributions.

On moving into the weaker field met at the downstream crossing point group **1**, now outbound, will be confined to the same smaller range of pitch angles that they occupied in the upstream ramp, and to a still smaller range in the outbound part of the downstream solar wind.

An equivalent process takes place starting from the downstream crossing point, with group **6** being mirrored to **6′**, and **5** to **5′**. Only the smallest-pitch-angle electrons, **4**, reach the magnetosheath and proceed to the upstream ramp and then on into the outbound portion of the solar wind.

Note that, since magnetic mirroring serves to mirror just the right number of particles to compensate for the narrowing of a tube of force, velocity-space densities will be the same everywhere as long as the initial distributions **1–6** were the same. The distortions undergone by the magnetic field as the result of the shock will therefore, in purely adiabatic motion, be completely transparent to the electron distribution as a whole. The factors leading to this state of affairs will continue to act when other influences come into play.

Another point to bear in mind is that transitions observed from a moving spacecraft do not necessarily correspond to transitions existing in space. For a start, spacecraft are not, in general, travelling exactly along the route along which the electrons are guided, i.e. along the direction of the magnetic field, which is in any case different on the two sides of the shock. For this reason, the electrons seen in the magnetosheath will have arrived from a neighbouring, generally different, region of the solar wind than that actually encountered. So, any assumption of equivalence will be invalid unless the solar wind is uniform over the distances involved. Although there is usually no direct means of assessing whether there were significant spatial variations, there is, in the case of the particular crossing we are studying, the circumstantial evidence from the relatively constant distribution found in the solar wind subsequent to the crossing that any such variations are likely to have been insignificant.

A distinct possibility, though, that does give need for caution is that of a change in solar-wind properties actually being responsible for the fact that the spacecraft suddenly found itself transported from the magnetosheath into the solar wind. We know from the scatter of points in figure 4.2 that the position of the bow shock varies considerably in response to changes in the solar wind. If a change in solar-wind properties had been responsible for the bow shock moving inwards, it would follow that the solar wind the spacecraft found itself in after the transition would have been different to that prevailing while it was in the magnetosheath. Again, this possibility cannot be dismissed, but the circumstantial evidence of the steady distributions found in both the magnetosheath and the solar wind lends some support to a working assumption of there being little significant temporal change on this occasion. The very mention here of temporal change should alert us to the possibility of acceleration and other effects arising from electric fields induced by changes in magnetic fields.

Many attempts have been made to account for the electron acceleration and other effects that occur in the regions of bow shocks. These fall into two distinct and fundamentally different categories. The first, which is dominant in the literature, is a category appealing to static and unchanging potential differences of one form or another. Bearing in mind our earlier finding that in order to cause acceleration the accelerator must itself undergo a change, we should be especially wary of any such theory from the outset, but there is no harm in looking at what is on offer. We do not have to buy.

4.3   Potential-difference theories

We discuss three mechanisms under this category. The first exploits a potential difference believed to exist across the shock layer. The second invokes the interplanetary electric field and the third combines the two.

4.3.1   Cross-shock potential difference

One-dimensional analyses of shocks in which the shock layers are considered to be planar, or laminar, sheets of infinite lateral dimensions, require there to be a potential difference across the shock. The potential difference is of a magnitude and direction to retard positive ions, and to accelerate electrons into the downstream region.

A question we must ask is whether the one-dimensional analysis is valid in this respect. Any potential differences that occur in the region of the shock must originate from charge imbalance. Since any real region of charge excess will necessarily be finite in size (it cannot be infinite, as tacitly assumed in one-dimensional analysis), so we can be certain that at distances suitably large compared with those of the space-charge region, the potential disturbance will be negligible. Potential jumps cannot extend indefinitely downstream, separating the universe into two halves. Instead, space-charge imbalance will lead to localised potential wells, positive and negative, with duly closed equipotential surfaces. The question then becomes, how localised are these potential wells? The answer to this depends on the actual deployment of the charge. Deployment has, as far as I am aware, never been specified, and the equipotentials relating to the supposed cross-shock potential are left as unphysical infinite planes. The misleading conclusions drawn from another one-dimensional analysis, that of double layers, which led falsely to the conclusion that there would be a net potential difference across such structures, will be recalled here. I am unaware of any direct measurements of, or attempts to obtain direct measurements of, the envisaged cross-shock potential.

Our concerns about the inevitable return to the original potential can be held in abeyance in what follows if we restrict ourselves to considering only the immediate vicinity of the shock, and consider the region of the magnetosheath in which the measurements were made to constitute a positive potential well into which electrons will be drawn.

Figure 4.10 summarises the basic elements of the first cross-shock potential theory [176, 183]. Electrons of the solar wind **1** are accelerated by a potential difference Φ assumed to exist between the solar wind and the magnetosheath. Acceleration increases the energies of all electrons by the same amount ($e\Phi$) and the distribution is shifted bodily to **2**, as discussed in Chapter 1. Distribution **2** with its steep positive slope is unstable to a re-distribution of energy, and plasma waves are generated and serve to increase entropy. The result is a flatter, broader distribution **3**, which is this model's prediction for the magnetosheath. The model envisages the magnetosheath distribution as being formed, possibly, from a series of such acceleration/re-distribution stages.

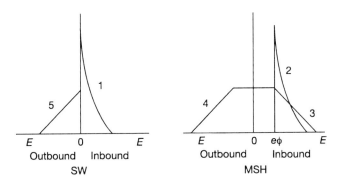

Figure 4.10. Schematic of potential-difference theory of electron acceleration at the bow shock. Electrons **1** of the solar wind (left) gain energy from an invoked potential difference to become the accelerated distribution **2** in the magnetosheath (right). The distribution is unstable and broadens to form distribution **3** which is magnetically mirrored, transported from the conjugate shock crossing or scattered back to form an outbound group **4**. On traversing the potential difference, now in the opposite sense, the electrons are retarded to form distribution **5** (from [176]).

Electrons **4** arriving from the conjugate portion of the shock, or being scattered, travel back into the solar wind, being retarded in the process, to return to the solar wind as distribution **5**.

This model accounts for several of the common key features. Its magnetosheath distribution **3** is broader than the solar-wind distribution **1**, and the predicted flat-topped version is certainly seen—after measurement adjustment and extrapolation—on occasions. The model could readily accommodate the rising-shelf distributions if erosion of the peak in **2** had not proceeded to completion. The high-energy electrons emerging into the solar wind **5** are also accounted for.

There are a number of factors, though, on the debit side. The velocity-space density reduction which is central to the theory results in a depression of the emerging distribution **5** relative to the incident distribution **1** at low energies. Observations show no such depression. The theory does not give rise to the falling-shelf distributions sometimes seen in the magnetosheath (e.g. figure 4.6), since erosion will not proceed beyond the stage of levelling.

4.3.2 Interplanetary electric field

Figure 4.11 sets the scene for this model [184]. Here, the cross-shock potential has been dispensed with, at least as a key controlling factor. Electrons gyrate about the magnetic field and drift in the solar-wind electric and magnetic fields encountered in the region of the shock. In this simplified picture we consider the magnetic field to jump from a low upstream value to a considerably higher value downstream in the magnetosheath. The reduced gyro-radii on the downstream side cause electrons to gradient drift with a component along the shock surface. Since this motion produces also a drift across electrostatic equipotentials, the electron's kinetic energy will change. Due to the means by which they are set up in the solar wind, the equipotentials will inevitably be marshalled so that the electron gains energy. The details of the fields and the initial conditions determine whether particular electrons or ions succeed in crossing the shock, as in the figure, or whether they find themselves after acceleration being directed back into the solar wind. If they do cross the shock they will '$\mathcal{E}$-cross-B' drift with no further change in mean energy.

This acceleration process is known as 'gradient-drift' or 'shock-drift' acceleration, the latter term being inclined to be reserved for situations where the particle makes many gyrations before escaping upstream or downstream. Electrons at the Earth's bow shock are envisaged by the model's proponents as making only one gyration before escaping, and, for this, the former term is preferred.

The events depicted in figure 4.11, with an electron 'starting' within the system of equipotentials, will cause electron energies to increase approximately in proportion to magnetic field strength, leaving the shape of the distribution (on a $\log f - \log E$ plot) essentially unchanged. The increase will be predominantly in the perpendicular motion. The process is found from detailed analysis to depend critically on the angle between the magnetic field and the shock front, being limited to where these are parallel to within just a few degrees, i.e. to almost exactly perpendicular shocks.

These predictions are at variance with observations of the change from the near power-law of the solar wind to the broader 'flat-topped' or 'falling-shelf' distributions of the magnetosheath. They do not account for electron energisation taking place at shocks that are only quasi-perpendicular, such as the one chosen for our case study. There is a slight relaxation in the perpendicularity requirement if pitch-angle scattering is included in the model, but the energisation is correspondingly weaker. If these inconsistencies can

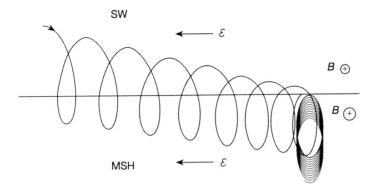

Figure 4.11. The 'shock-drift' theory of acceleration. The path shown is that followed by an electron subject to a jump in magnetic field strength and a tangential electric field at the bow shock. The guiding centre drifts to the right as the result of the magnetic field jump, and, in the process, the electron becomes energised by the electric field. The guiding centre is also drawn towards the region of stronger magnetic field, until, eventually, it is 'captured' by it. Subsequent motion is an '$\mathcal{E}$-cross-B' drift downwards. For the purposes of presentation the horizontal axis has been compressed. The net effect is that the electron gains energy by migrating towards an electrostatic potential more positive than that from which it 'started' (based on [184]).

be overlooked, the positive claim can be made, following a more detailed analysis, that the energisation would be expected to reach a peak within the shock layer. This occurs at the point of maximum magnetic field strength, i.e. at the 'overshoot' often observed in the magnetic profile.

4.3.3 Cross-shock potential difference and interplanetary electric field combined

Figure 4.12 summarises the main features of the potential-difference acceleration model which contrives to accelerate electrons by just a fraction of the (assumed) cross-shock potential [185]. The curved lines representing the equipotentials considered to prevail in the region of the shock are the result of a superposition of a cross-shock potential difference and potential differences arising from solar-wind flow. Solar-wind ions are considered to cross the shock layer undeflected, to experience and be retarded by the full cross-shock potential difference. Electrons, on the other hand, are, because of their lower magnetic rigidity, guided by the magnetic field along paths that cross fewer equipotentials.

In this model the observed broadening of the electron distribution is attributed solely to the energy-dependent motion in the combined electric and magnetic fields in the shock layer.

As with the previous model, the full implications cannot be assessed until

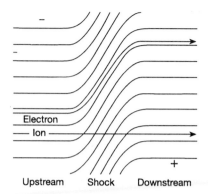

Figure 4.12. Schematic diagram of the bow-shock acceleration process in which solar-wind and cross-shock electric fields combine to accelerate electrons through a relatively small change of electrostatic potential, while ions are retarded by a larger change (from [185]).

the equipotentials are defined as closed surfaces. Such a step would inevitably show that the overall system was inert and that there would be no net change of energy for any species of particle.

Particle trajectories and energy gains and losses in this model are commonly described in a frame of reference (the de Hoffman–Teller frame [186]) in which the upstream and downstream electric fields vanish. This is purely a mathematical convenience (for some) and makes no difference to the physics as long as it is remembered to transfer back to the shock, or observer's frame when assessing the results.

4.3.4   Summary

These theories all suffer from being fundamentally incomplete. They all restrict themselves to considering motion of electrons in limited regions of static and unchanging electric and magnetic fields. Since such fields are conservative, it follows that, had motion in the fields not considered been included, the net effect would necessarily be zero. Another way of highlighting the same problem is to note that, although the electrons are seen as gaining energy, nothing seems to have lost energy!

4.4   Wave theories

This second category of theories consists of those based on wave–particle interactions. You will recognise at once that, with these two categories we are entering a dichotomy similar to that existing for the aurora.

4.4.1   Advent

It has long been known that the terrestrial bow shock abounds with electric and magnetic waves [187]. Some progress has been made in identifying the various types of wave that occur, but this is at present still far from complete. The possibility that solar-wind electrons might be accelerated by Landau resonance with lower-hybrid waves in the region was suggested by Papadopoulos in 1981 [188]. This possibility was also mooted by Vaisberg and colleagues in 1983 [187] and by Galeev and colleagues in 1986 [189]. Our first involvement in 1985 [83] arose directly from the striking similarity in our own measurements of the solar-wind to magnetosheath transition at the bow shock with the magnetospheric to auroral electron transformation with which we had previously been concerned (compare, for example plates 6(*b*) and 2(*a*)). The fact that the electron energies affected were in both cases of the order of the random or thermal energy of the ions—a few keV in the aurora, and a few hundred eV at the bow shock—added fuel to this proposition. The ever present scientific goal of explaining more with the same was a further driving force. This section introduces and shows the current state of development of a model of lower-hybrid wave acceleration at the bow shock based on the same principles as the auroral wave model.

4.4.2   Lower-hybrid waves

In a plasma where the electron or ion plasma frequency greatly exceeds the corresponding gyro-frequency, as is the case in the solar wind and the magnetosheath, the lower-hybrid resonance frequency becomes, as we saw in the previous chapter, the 'geometric mean' of the electron and ion gyro-frequencies, i.e.

$$(\omega_{\text{lh}})_{\text{dense}} \sim \sqrt{\omega_{\text{ce}}\omega_{\text{ci}}}.$$

If the magnetic field strength is B nT, the lower-hybrid resonance frequency ν_{lh} ($= \omega_{\text{lh}}/2\pi$) is given by

$$\nu_{\text{lh}} \sim 6.5 \times 10^8 B \ \text{Hz}.$$

During the transition from the solar wind to the magnetosheath, the lower-hybrid frequency rises in our case study from just below 10 Hz to a few tens of Hz. Lower-hybrid waves would be expected, therefore, to lie in a frequency band extending upwards from these frequencies to about 100 Hz. Such frequencies occur towards the lower limit of measurements currently available. The measurements also show consistently that the highest wave power per unit frequency range—or power spectral density—is found at the experiment's low-frequency cut-off of around 10 or 30 Hz. It should be pointed out, though, that these observed waves have been interpreted as being electromagnetic waves of the type known as whistlers. There has been no assessment, to my knowledge, of what proportion of the observed electric oscillations might be lower-hybrid waves.

The velocity of lower-hybrid waves in directions perpendicular to the magnetic field in the shock layer is, as in the case of the auroral acceleration region, similar to the random or thermal velocity of the heated solar-wind ions. Since thermal energies in this region are of the order of 100 eV, the perpendicular velocity is $\sim$ 150 km s^{-1}. The wavelength in this direction is this velocity divided by the frequency of, say, 30 Hz, i.e.

$$\lambda_\perp \sim 5 \text{ km.}$$

Due to the angle of propagation the wavelength parallel to the magnetic field is greater than this by the familiar factor of $\sim$ 43, i.e.

$$\lambda_\parallel \sim 200 \text{ km.}$$

The parallel phase velocity is this parallel wavelength multiplied by the frequency, or

$$v_{\phi\parallel} \sim 6.0 \times 10^6 \text{ m s}^{-1}.$$

This is the velocity of an electron of 100 eV, re-affirming the linkage effected by lower-hybrid waves between the parallel motion of electrons and the perpendicular motion of ions of the same energy.

It should be noted that the above estimate makes $\lambda_\parallel$ comparable to the estimated thickness of the shock layer itself.

4.4.3 A wave model

Work has begun only recently on a model of stochastic acceleration by lower-hybrid wavepackets. The auroral model itself is not immediately applicable because it would deliver structure in the angular distributions (e.g. beams, conics etc) not seen at the bow shock or in the electrons travelling upstream from the shock. In acknowledgment of the generally fairly isotropic distributions found in the shock environment, we have added to the model a degree of angular scattering. The effect of the scattering is, as would be expected, to smooth out any latent angular structure. It also presents a semi-permeable barrier to progress through the shock layer in either direction, thus further weakening, as required, the coherence between velocity-space densities in different parts of the region.

The result of a preliminary test of the model is shown in figure 4.13. The figure shows that a magnetosheath distribution can be developed without this same distribution returning to the solar wind. The electrons that do return are in accord with observation in having an enhanced high-energy tail, but at present the returning flux is too great. These early results are therefore no more than encouraging. Wave amplitudes used in the analysis are $\sim$ 2.5 V, with corresponding electric fields of $\sim$ 1 mV m^{-1}.

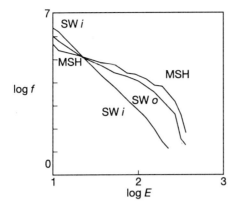

Figure 4.13. First results of an application to the bow shock of the wave model of auroral electron acceleration, modified by the inclusion of directional scattering. The model produces a magnetosheath distribution similar to the observed one in figure 4.2. The electrons returned to the solar wind (**SW** *o*) are qualitatively similar to those of figure 4.7 in being of higher velocity-space density than the ingoing ones (**SW** *i*) at high energy, and lower at low energy. They also lie below the magnetosheath ones at high energies, as observed. Quantitatively, though, the outbound electrons depart too far from the ingoing ones for the model to be considered more than promising at this stage.

4.4.3.1 Energetics

Among the various quantities that can be calculated from the velocity-space distributions of figures 4.5, 4.6 and 4.7, are the amounts of energy carried by the electrons into and out of each unit area of the shock layer every second. The contribution to this energy flux made by electrons in a velocity range dv centred on v is, assuming isotropy over each of the inbound and outbound directions,

$$dP = 1.4 \times 10^{-48} f v^5 \, dv \; \text{W m}^{-2}$$

where f is in km^{-6} s^3.

When the contributions from the full range of observed energies or velocities are summed we find that the net flow at the solar-wind surface of the shock layer is

$$P_{\text{layer to SW}} = 0.03 \; \text{mW m}^{-2}$$

and that at the inner surface

$$P_{\text{layer to MSH}} = 0.12 \; \text{mW m}^{-2}.$$

In order to find the net rate of input of energy to an element of the shock layer, having, say, unit area at the outer surface, it is necessary to allow for the fact that the area of the tube of force tapers towards the magnetosheath. This tapering corresponds exactly to the increase in magnetic field strength. Since the field strength at the inner surface was approximately six times the field at the outer surface, the area at the inner surface is six times less than that at the outer surface. The net rate, then, at which electrons remove energy from an element of the shock layer having an area of 1 m^2 at the outer surface is equal to the outer area multiplied by the former of the above energy fluxes plus one-sixth of the latter, or

$$P = 0.05 \text{ mW m}^{-2}.$$

This applies regardless of the energising mechanism.

In the wave theory this power is drawn from lower-hybrid waves populating the layer. For the mechanism to be viable, therefore, the waves must be replenished from a source delivering energy at the above rate. The solar wind is the obvious candidate for the power source. In the case study the solar-wind ions, predominantly protons, had velocities of 400 km s^{-1}, corresponding to kinetic energies of just below 1 keV. Their number density was approximately 10^7 m^{-3}. These combine to give an incident energy flux of

$$\mathcal{P}_{\text{SW in}} \sim 0.6 \text{ mW m}^{-2}.$$

On leaving the shock layer for the magnetosheath the velocities were reduced to 120 km s^{-1}, corresponding to a mean kinetic energy of about 80 eV. The number density had increased to approximately 10^8 m^{-3}, to make the energy flux leaving the layer for the magnetosheath

$$\mathcal{P}_{\text{SW out}} \sim 0.2 \text{ mW m}^{-2}.$$

A substantial quantity of the energy incident upon each square metre of the shock surface was clearly lost each second in the shock layer (around 0.4 mW), a quantity far exceeding the 0.05 mW required, according to the above estimate, to energise the electrons.

No claim is made for these numbers to be at all typical, and indeed the uncertainties involved and approximations adopted are too great for the figures to represent an energy budget. They do however illustrate some of the principles involved and serve to indicate that the solar-wind ions are a credible source of energy for the replenishment of the waves that accelerate the electrons. One of the key outstanding issues is the exact nature of the wave activity at the shock.

The mechanism by which ion beams give rise to wave turbulence is outside the scope of the present work. However reassurance that wave turbulence is created by ion beams may be gained from the fact that the process is essentially the same as that exploited in tokamak plasmas [150].

4.5   Conclusions

The situation is similar to that of the aurora. All models based on potential differences appear to contain a common fundamental flaw. Lower-hybrid waves seem in principle to be capable of producing what is observed, though further details of the waves and further development of models based on them are required before any firm conclusions can be drawn.

Chapter 5

Electron acceleration in the neighbourhood of 'artificial comets'

5.1 'Artificial comets'

On 27 December 1984 an unprecedented sighting was made from Boulder, Colorado. High in the dawn sky, a comet-like object, brightest at the head and with a tail some 2° in angular 'length'—approximately four times the angular diameter of the Sun or the moon—made a sudden and fleeting appearance. The phenomenon lasted for about four minutes before becoming too faint to be seen even with a low-light television camera and other extremely sensitive optical equipment that had recorded its brief life.

The object was an 'artificial comet' [190, 191] produced more than 100 000 km above the Earth's surface in the solar wind as one of the highlights of the AMPTE mission [177, 178]. In the previous chapter we saw some observations of the Earth's bow shock made by the UKS spacecraft of this mission, here we shall explore some phenomena induced by injection of foreign material into the solar wind and magnetosphere by another of the mission's spacecraft, the Ion Release Module (IRM). Our pre-occupation will be with the electron acceleration promoted by these 'active' experiments.

5.1.1 Creation

The IRM had the remarkable ability to release on command from the ground clouds of ions into the solar wind, the magnetosheath and the terrestrial magnetosphere out to distances of 18.5 R_E. The aims were twofold. One was to explore, together with its travelling companion the UKS, local and temporary perturbations caused by the ions. The other was to try to discover, working in conjunction with the third of the mission's spacecraft, the Charge Composition Explorer (CCE), patrolling within the magnetosphere, how solar-wind material enters the magnetosphere, using the ions as tracers. The ions used were singly ionised lithium and barium.

5.1.1.1   Expanding shells of atoms

Atoms of lithium and barium both form explosive mixtures with molecules of copper oxide. The chemical reactions are

$$2Li + CuO \rightarrow LiO + Cu + Li$$

and

$$2Ba + CuO \rightarrow BaO + Cu + Ba.$$

Products of these interactions are lithium and barium atoms flying away from the seats of the explosions in shells expanding (see figure 5.1) at ~ 3.7 km s^{-1} and 1.4 km s^{-1}, respectively. In the presence of solar ultraviolet light, the atoms become ionised, not all at once but with time constants of ~ 1 hr in the case of lithium and a much shorter ~ 28 s in the case of barium, with the effect that approximately two-thirds of the atoms remaining at any time become ionised during each of these characteristic periods.

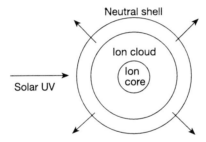

Figure 5.1. Basic processes of an 'ion release' experiment. An expanding shell of neutral particles becomes ionised by solar uv radiation to create a diamagnetic ion core, a surrounding ion cloud and an expanding neutral shell (see [192]).

5.1.1.2   Ion clouds

As one of these shells of atoms expands it deposits an expanding plasma of ions and their former associated electrons. The number density of the plasma is at its greatest at the centre and at the beginning of the process because the number density of atoms is greatest then.

5.1.2   Magnetic cavities

Since the explosions took place in a medium containing a magnetic field, currents were induced in the expanding plasma in the sense to keep the magnetic flux threading the plasma cloud constant. The combination of constant total flux and increasing size naturally leads to a reduction in magnetic field strength within the cloud and an enhanced field outside, as indicated at the upstream side of the ion 'core' in figure 5.2. The process recalls that by which the solar wind creates the 'garden hose' interplanetary field and the enhanced magnetic field upstream from the Earth's magnetosphere. Since the plasma has only a limited ability to set up currents, such domination of the magnetic field cannot continue indefinitely. A stage is reached when further expansion is stopped and the 'magnetic cavity' and the enhancement outside both reach their peak.

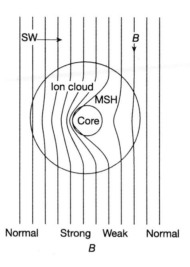

Figure 5.2. The magnetic field in the vicinity of an ion release in the solar wind. The field in the upstream part of the ion cloud is enhanced relative to the unperturbed solar wind. Field strength maximises at the edge of the ion core within which there may be an extremely weak field, or even a 'magnetic cavity' (see [192]).

The motion of the surrounding plasma also plays a part in the competition for territory, helping to resist expansion upstream but facilitating expansion downstream, as also indicated in figure 5.2. The process actually mimics in these respects the interaction of the solar wind with the atmospheres of unmagnetised planets and with the gas, ion and dust environments of comets.

5.1.3 'Pick-up'

Figure 5.3 illustrates the effect of the combined action of the interplanetary electric and magnetic fields on freshly created ions. An ion setting out in the electric field direction travels only a limited distance before this progress is arrested by deflection in the magnetic field and it is returned back to the same electric potential from which it started, but at some distance downstream. The sequence then, naturally, starts again. When the process is examined quantitatively it turns out that the particle completes a cycloidal path in the same time that the solar wind takes to travel in a straight line between the points of the cycloid. The particle is, therefore, making progress, albeit somewhat erratically, in the same direction as the solar wind and at the same average velocity. The motion is the same as that of a point at the rim of a rotating wheel. The particle's velocity parallel to the solar wind ranges from zero at the points of the cycloid, to twice the solar-wind velocity half-way between. A particle joining the solar wind in this way is said to be 'picked-up' by the solar wind. In the general case of the magnetic field not being exactly at right angles to the flow, the process still works but the pick-up velocity is then less than the solar-wind velocity. Only in the case of the field being parallel to the flow is there no pick-up. Due to the fact that the directions of both the electric and magnetic forces reverse for particles carrying different signs of charge, electrons and ions become picked-up to flow in the same direction. It will be apparent that this pick-up process is a special case of motion in crossed electric and magnetic fields discussed in Chapter 1. This is the same process as that producing the co-rotation discussed in Chapter 2.

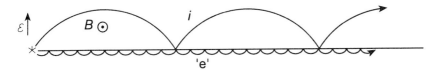

Figure 5.3. Electron and ion 'pick-up'. Following deposition (*) by ionisation within a medium of crossed electric and magnetic fields, electrons and ions undergo '$\mathcal{E}$-cross-B' drift. If their velocities on release are very much less than their ultimate drift velocity, the electron and ion execute simple cycloids. For convenience of illustration the 'electron' 'e' is given a mass equal to one tenth of the ion mass.

The fact that energy has been imparted to charged particles by static and unchanging electric and magnetic fields might seem at first to contradict our earlier conclusion that the conservative nature of such fields makes this impossible. However, there are new elements here—the acts of injection and photo-ionisation—which together provide the essential irreversible, non-conservative element. This fact is a telling comment on a profound difference between gravitational and electrostatic forces. There is no gravitational equivalent of a particle that is repelled by a concentration of charge. If there were, we could travel with it freely to the top of a hill and release it to fly away into space while we skied down! In the pick-up process, the ion and electron travel joined and unhindered to the electric potential from which they are separated by photo-ionisation, whereupon, by virtue of their opposite sign of charge they both make 'downhill' excursions in opposite directions, as though both performing frictionless skiing. The average gain of kinetic energy corresponds to the average potential energy lost which, in turn, corresponds to the average distance of excursion.

Where has the energy come from? Since acceleration is inextricably linked with net movement across electric equipotentials, in the sense of neutralising the electric field set up by the solar wind, the energy could be thought of as having been derived from a reduction of this field and the energy associated with it. Should this begin to happen, however, the solar wind would be deflected to replenish the space charge it had set up originally to allow force-free flow. The energy required for restoring the electric field would therefore derive from the solar wind. In order to accomplish this, solar-wind ions would need to have a component of flow in the opposite direction to that in which the injected ions were displaced during pick-up. This deflection may be seen as a recoil of the solar-wind ions to the displacement of the injected ions. The recoil in the opposite direction to the displacement of the electrons would be very small in comparison because the momentum of the electrons is much less than that of the injected ions. In the course of attempting to restore the status quo, the ions of the solar wind lose an energy equivalent to that gained by the picked-up ions. The overall effect of the injection is, therefore, for the solar wind to be slowed down by the 'load' of fresh ions it has to carry. The whole process may therefore be seen as an elaborate, collisionless version of the simple case of one mass 'knocking on' another.

5.2   The AMPTE lithium experiments

On 11 September and 20 September 1984, 10^{25} atoms were released from the IRM into the solar wind upstream from the Earth's bow shock [193]. Due to the relatively long ionisation time for lithium, most of the resulting ions can be assumed to have created a giant cloud comparable in size to the Earth itself. This was not observed directly. Indeed any attempt would have been fruitless since the Sun was directly behind the cloud as seen from the Earth. Number densities in the clouds may be estimated, from the expansion velocity (3.7 km s^{-1}) and the time constant for ionisation (1 hr), to have matched the solar-wind number densities of 10^6 m^{-3} in the first event and 10^7 m^{-3} in the second, at distances as far as 1000 km from the point of release, and to constitute a significant perturbation well beyond this.

The intention was for these ions to be picked-up by the solar wind and carried forward into the magnetosphere as tracers of solar-wind entry. Unfortunately, the interplanetary magnetic field suddenly became almost radial during the first event, with the result that pick-up would have been rather ineffective. The field was, however approximately perpendicular during the second event so the ions should have been picked-up then. On neither occasion were any of the ions detected by the CCE waiting in the magnetosphere. The negative result, readily accountable for only in the former instance, poses serious questions for theories of solar-wind entry into the magnetosphere [193].

Our concern here, though, is not with the giant ion clouds but with the immediate environments of the magnetic cavities created around the release points. The IRM, naturally at the centre of the disturbances, and the UKS, just

33 km away on both occasions, were ideally placed to see what went on. The IRM confirmed that magnetic cavities were created, the field falling below the lowest measurable on both occasions. The UKS, just 2 km outside the first cavity saw the field strength drop to less than half its pre-release value. The UKS was 8 km outside in the second event and saw no decrease. On both occasions the enhanced field regions upstream from the magnetic cavities were transported to and past the spacecraft as the solar wind drove the plasma and the cavity within it downstream. This motion drew both spacecraft through the magnetic field enhancements that had formed just upstream. At their maxima, the field enhancements were approximately four and six times the pre-release values. After 1 min in the first event and 30 s in the second the disturbance had either subsided or had been carried so far downstream that the spacecraft found themselves again in the undisturbed solar wind.

Now to the electron measurements. Electron intensities at the UKS showed no significant departure from their pre-release (solar-wind) values until the enhanced-field region was reached. Electron velocity distributions taken at the time of greatest perturbations are compared with the solar-wind electrons in figures 5.4 and 5.5. If the perturbations are seen as increases in energy at constant velocity-space density the greatest factor of increase in the first event ($\times 4$) occurred for an initial energy of 30 eV, while the greatest increment (160 eV) occurred for an initial energy of 500 eV. In the second event the greatest factor ($\times 2.2$) occurred at 20 eV, and the greatest increment (80 eV) was at 500 eV. The perturbations represent, therefore, neither simple factors of increase, nor simple additions of energy. In this they are reminiscent of those seen in the aurora and at the bow shock. In common, too, with the bow shock, the distributions were approximately isotropic.

5.3   The AMPTE barium experiments

On 27 December 1984, 10^{25} barium atoms were released by the IRM into the solar wind approximately 100 000 km above the dawn side of the Earth [190, 191]. This location would normally lie well inside the magnetosheath, but the unusually high solar-wind velocity (540 km s^{-1}) on that day brought the bow shock further downstream than usual with the result that the release took place in the solar wind. Logistic reasons involving the spacecraft orbit and the observing sites, made the Christmas period ideal for the experiment. Christmas Day was excluded because of poor visibility at the observing sites which included Hawaii where there was a snow storm! The release was duly made one orbit later.

The relatively low expansion velocity of the released barium atoms (≈ 1 km s^{-1}) and the short time constant for ionisation (28 s) served to keep almost all of the ions together in a dense cloud which created a magnetic cavity some 100 km in radius. Emissions from the excited ions produced the coma at the head of a comet-like object seen from the ground. Only a very small proportion of atoms escaped to give rise to a diffuse ion cloud beyond the

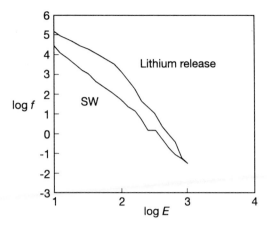

Figure 5.4. Electron energy distribution measured by UKS before and at the maximum of disturbance after the IRM release of lithium ions into the solar wind on 11 September 1984 (from [195]).

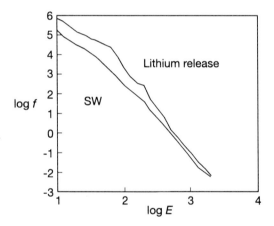

Figure 5.5. Electron energy distribution measured by UKS before and at the maximum of disturbance after the IRM release of lithium ions into the solar wind on 20 September 1984 (from [195]).

cavity.

Tracer ions were neither anticipated nor observed by the CCE from this experiment. Ions continuing to expand unhindered into the weak-field region downstream produced a distinctive tail, while ions at the boundary of the cavity were picked-up by the solar wind. Since the magnetic field happened to be almost exactly perpendicular to the solar-wind flow, and in the same sense as the field in figure 5.3, the picked-up ions were extracted northwards to create a rather jagged appearance of the north side of the coma and tail [190]. So far, everything fitted in with our (incomplete) understanding of what might happen. We expected solar-wind pressure to drive the whole object tail-first downstream. It did not—at least not at first; it moved southward, i.e. sideways! Why? (Cover up the next paragraph if you want to think about it.)

As discussed above, in order to displace the barium ions northward in the process of picking them up, the solar wind must recoil southwards. This causes the solar wind to impinge on the north side of the cavity so driving it southwards as observed. It will be immediately apparent that events will be made more complicated than this simple picture would suggest by the fact that localised change in solar-wind direction will produce localised changes in the pick-up process. Pick-up so rapidly eroded the plasma core in this experiment that the 'artificial comet' that had been expected to last for some hours actually disappeared from view after about 4 min. However, before it did so it was duly seen to set off downstream. The details of the physical processes by which solar-wind energy and momentum were imparted to the barium ions in the core and the tail of the 'artificial comet' remain something of a mystery. The fact that energy and momentum were imparted, as though in simple particle-to-particle collisions, provides another example of the way action at a distance in a collisionless plasma substitutes for collisional interactions such as friction and viscosity in denser media.

The magnetic cavity was approximately 70 km in radius on this occasion. The UKS some 170 km away from the IRM was thus outside the cavity, but at the edge of the coma. As the 'artificial comet' moved southward and then tailward the UKS and IRM found themselves, as they had in the lithium releases, in the enhanced magnetic field upstream and around the flanks of the object. Here, the electron intensities were far above their pre-release, solar-wind values. Plate 7(a) shows in spectrogram form the disturbance seen at the UKS. The velocity distribution taken at the maximum of the 4 min disturbance is compared with the pre-release solar-wind distribution in figure 5.6. Here the factor by which energy increases (again assuming velocity-space density to be conserved) is greatest ($\times 5$) for initial energies of 40 to 100 eV, while energy increments are greatest (400 eV) for an initial energy of 300 eV. Again, the perturbation can be described neither in terms of a uniform factor of energy increase nor by a uniform addition of energy.

A barium release was made into the magnetosheath on the evening flank of the Earth on 18 July 1985. This also produced an 'artificial comet' with a visible tail more than 1000 km in length. Unfortunately the UKS had ceased to operate by then, so direct comparison with the previous releases with the same instrument is not possible. The perturbation to magnetosheath electrons as observed from the IRM is shown in figure 5.7. The greatest factor of energy increase ($\times 6$) occurs at the lowest initial energy. The energy increment is at its greatest (180 eV) for initial energies between 180 eV and 500 eV. True to previous form, neither increments nor factors of increase are uniform.

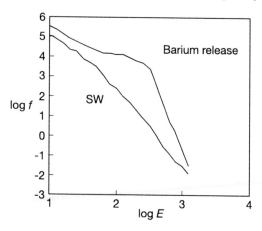

Figure 5.6. Electron energy distribution measured by UKS before and at the maximum of disturbance after the IRM release of barium ions into the solar wind on 27 December 1984 [83].

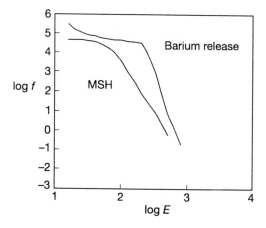

Figure 5.7. Electron energy distribution measured by IRM before and at the maximum of disturbance after the release from this spacecraft of barium ions into the magnetosheath on 18 July 1985 (courtesy G Paschmann).

An important additional item of information that may be gleaned from figure 5.7 is that velocity-space densities after the release exceed those within the energy range of the instrument in the magnetosheath. It follows that, if the former is an accelerated version of the latter, there must have been velocity-space densities in the magnetosheath higher than those of the plateau seen before the release. It seems, then, that the release has forced into the open the low-energy component of the magnetosheath often hidden from view, a concealment giving the false impression of a 'flat top' to some magnetosheath distributions.

5.4   Interpretation

The features of the perturbations of the electron distributions, common to all four active experiments described above are that (on the assumption that velocity-space density is conserved) energies increase by different amounts and different factors for different initial energies. Several theories have been considered.

5.4.1   Potential difference

The releases might, in principle, have produced positive potential wells in the regions where electron acceleration was observed. If this had occurred electrons would have been accelerated into the potential wells, gaining energy in the process. This is just the eventuality discussed in relation to the bow shock in the previous chapter, where a sharp peak would be expected to arise, before degenerating to leave a plateau.

The consistent finding of a negative slope rather than a positive slope or plateau argues against this interpretation here, just as it did in the case of the negative slopes observed in the magnetosheath and, indeed, in auroral rays. There is nothing in the electron measurements or any other measurements to suggest that a potential well has formed.

5.4.2   Adiabatic compression

Since these active experiments are essentially transient in nature it is important to consider the possibility of acceleration by induction. To this end we can estimate what the energy increase might be from the magnetic field enhancement while electrons were being guided along the tubes of force draped around the magnetic cavities.

The electrons we are considering have velocities of several thousand kilometres a second. This ensures that, even if they entered the enhanced field region with pitch angles large enough for them to mirror, they could not spend more than a very small fraction of a second in the region (say 0.1 s for the barium releases and ~ 0.01 s for the smaller of the two lithium disturbances). The changes in field strength observed over these times were very small. Even if all of the observed change were truly temporal, with no spatial component, the changes, and consequently the factors by which energies would be enhanced, would have been less than 1%. For this reason adiabatic compression seems unlikely to have played a significant role.

Another argument against this interpretation is that adiabatic compression would increase all energies by the same factor, and this was not what happened.

5.4.3 Waves

The preference shown by the accelerator for particular initial energies, whether for greater factors of increase or greater increments proclaims here, as it does in the aurora and at the bow shock, that the process involves resonance. We have, therefore, to ask whether acceleration could again be effected by resonance with waves.

In this connection it is unfortunate that acceleration takes place at such low energies that we cannot see what prevailed before at the crucial high velocity-space densities. The action is at too low an energy even to allow us to confirm whether or not the increases in velocity-space densities at higher energies are compensated by the expected reductions at low energy.

One positive indication that waves might be involved is the fact that the regions where the accelerated electrons were observed—immediately upstream from the obstacles presented by the ion cores—exhibited wave activity many orders of magnitude greater than in the uninterrupted solar wind. In fact, the electrostatic waves resulting from the ion releases were even stronger than those normally encountered at the bow shock [196].

Uncertainties over the full picture of the strength and extent of the waves precludes a detailed numerical assessment of the amount of acceleration the waves might be capable of imparting. We know enough, though, to keep the possibility alive.

The rms electric field observed by the IRM, thought to have passed closer to the heart of the disturbance than the UKS, though not necessarily through the maximum wave activity, was at times ~ 50 mV m^{-1}. The waves were electrostatic and within the lower-hybrid range of frequencies which here extended upwards from 14 and 28 Hz in the lithium events and from 80 Hz in the barium releases. This is one-quarter of the field estimated in Chapter 3 to be capable, if it extended over a distance of 10 000 km, of producing a peak at 10 keV. The distances involved in the releases are likely to be much smaller than in the aurora. When we note that the UKS, located between a cavity radius and a cavity diameter away, saw a much lower field, the distance over which the waves occurred is unlikely to have been greater than 100 km in any of the events. The earlier finding that field strength necessary to perform a given modification varies inversely with the square root of the length of the acceleration region, suggests that the reduced scale size would increase the field strength required by a factor of ten or more. However, the energies involved in the releases, being less than 100 eV, should to a first approximation compensate for the reduced scale size. There are several other factors which would need to be taken into account before knowing whether the remaining factor of four (in our estimate) can be accounted for. The most important of these, as our analysis of the aurora showed, are the amplitudes of the wavepackets (a few large-amplitude wavepackets being more effective at acceleration than many small-amplitude wavepackets).

5.5   Conclusion

Stochastic acceleration by lower-hybrid waves appears to be a very strong possibility [83] and, at present, the only viable one on offer.

Chapter 6

Electron acceleration in solar-wind events

Alert observers have noticed that spacecraft sampling the solar wind upstream from the Earth occasionally and briefly become engulfed in what seems to be an alien plasma [197]. Everything seems to shape up for an excursion into the magnetosheath, but events then take a turn for the unexpected. These 'solar-wind events' display an aspect of electron acceleration that we have not come across before in these studies. They are, therefore, of great prospective value.

6.1 Observations

6.1.1 Particles and fields

6.1.1.1 Electrons

Solar-wind events are particularly well revealed in electron energy–time spectrograms. Plate 7(b) shows an event occurring shortly after the UKS emerged from the magnetosheath to the solar wind on 30 October 1984. Initially intensities rise from their solar-wind values to emulate those found in the bow-shock ramp and the distribution broadens, as if re-entering the ramp en route for the magnetosheath. However, the transition is to a distribution not seen before. At low energies intensities revert to values similar to those of the solar wind, while high energies reach intensities even higher than those in the magnetosheath. After a period of about 2 min, which seems to be the norm, the ramp-like distribution is encountered again on the way back into the solar wind. For convenience we refer to the central region as the 'core' of the phenomenon, and the ramp-like features on either side as the 'wings'.

6.1.1.2 Ions

Solar-wind ions provide further crucial information. In common with the electrons, the ions have their velocity distribution broadened on travelling through the 'wings' to the 'core', as though entering the magnetosheath [197].

196

In the core, just as in the magnetosheath, ions have lower velocities than in the solar wind, and are deflected away from their previous near-radial direction. Their number densities in the core, however, which necessarily match electron densities found there, are much lower than those normally encountered in the magnetosheath. Another unusual feature is that some of the ions found there are singly ionised oxygen, an ion more at home in the Earth's upper atmosphere and magnetosphere than in the solar wind.

6.1.1.3 Magnetic field

The magnetic field is similarly deceptive. Its strength rises in both wings above solar-wind values as though on a transition between the solar wind and magnetosheath, but in the core takes on solar-wind values, often exhibiting great fluctuations [197].

The events are most commonly seen when the interplanetary magnetic field is in the near radial direction [198], thus when there is a connecting channel between the spacecraft and the Earth's bow shock. The conjugate part of the shock under these circumstances is necessarily a parallel or quasi-parallel shock. After the event has passed the field is usually set at a considerable angle to the pre-event, near-radial direction.

6.1.1.4 Waves

Electrostatic waves are a consistent feature of the 'wings' of the events [197]. Due largely to the fleeting nature of the phenomenon, there has not yet been a thorough survey of wave modes and amplitudes.

6.1.2 Locations

The arrowheads in figure 6.1 give the locations of the full 19 solar-wind events seen by the IRM over six months of exploration of the upstream solar wind [198]. The direction of the arrowheads indicates the direction of ion flow within the cores of the events. It is very noticeable that, with just three exceptions—all close to local noon and in two of which the direction changed—the directions are, like those of the magnetosheath, in the sense to avoid a downstream obstacle.

Another noticeable point is that the events tend to cluster just upstream from the typical position of the bow shock, as indicated by the shock positions transported from figure 4.2. The clustering is especially significant in view of the fact that, if the events occurred randomly in space, most would be seen near the satellite apogee of 18.6 R_E where it spends more time than at any other distance.

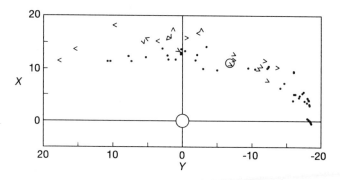

Figure 6.1. Locations of solar-wind events, as detected by AMPTE-IRM. Coordinates are as for figure 4.2. The arrowheads show the direction of the deflected solar wind (courtesy G Paschmann). Repeated for reference are the locations of the bow shock determined by the UKS during the same (late 1984) epoch. The location of the 30 October event, used for the case study, is ringed.

6.2 A case study

With the above general picture in mind let us undertake a case study of the event depicted in plate 7(*b*)—that of 30 October 1984.

On this occasion the UKS and IRM were located approximately 14 R_E from the Earth at a local time of 0930 h, at the position circled in figure 6.1. They had crossed the bow shock just over an hour earlier. The magnetic field was directed at $\sim 300°$ to intersect the bow shock with almost exact parallel-shock geometry. True to form for parallel shocks, the crossings had been quite gradual and it was not very obvious when the crossing had been completed. In this situation, a minor irregularity in the measurement of any quantity could easily have been dismissed as a brief re-entry into the magnetosheath due to an outward excursion of the bow shock. But Trefor Edwards, watching the real-time displays of UKS measurements, noticed that something very different was taking place.

6.2.1 Electrons

Electron intensities at three representative energies are shown in figure 6.2 for a four-minute period centered on the event, along with the simultaneously measured magnetic field magnitude. The first and last minutes show typical solar-wind values for both electrons and the magnetic field. Close inspection shows that, although the electrons return to their original intensities, the magnetic field does not quite do so. During the central minute, i.e. within what might be termed the 'core', intensities at the lowest energy are depressed relative to the solar wind, while those at the two highest energies are enhanced. The magnetic field is highly irregular, but not noticeably different on average to the solar-wind field. The half-minute transitions, or 'wings', show intensities that are enhanced at all energies relative to the solar wind. At the two lower energies the intensities are much higher than in either the solar wind or the core. At the highest energy the intensities match the highest encountered in the core. The magnetic field strength in the wings is somewhat irregular, but on average roughly a factor of two higher than in the solar wind. Electron energy distributions in these three regions are shown in figure 6.3.

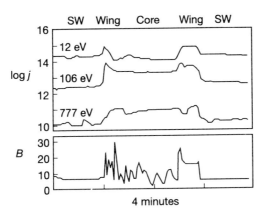

Figure 6.2. Electron intensities at three selected energies, and magnetic field strength at UKS during the solar-wind event of 30 October 1984. The perturbation regions of the 'wings' and a 'core' are discussed in the text (adapted from [197]).

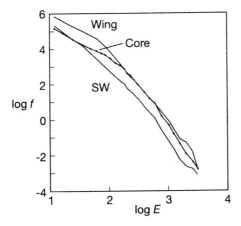

Figure 6.3. Electron energy distributions in the solar wind, the core and the wings of the case-study event of 30 October 1984 [47].

6.2.1.1   Solar wind

The solar-wind electrons are well represented by a kappa distribution. The electrons are almost isotropic. There is, however, at all energies greater than 50 eV a slight tendency towards counter-streaming parallel and anti-parallel to the magnetic field.

6.2.1.2   Wings

The 'wings' of the event display, relative to the solar wind, higher velocity-space densities at all (observed) energies or, equivalently, higher energies at all given velocity-space densities. The transition is equivalent to neither uniform addition of energy, nor a uniform factor of increase. The energy increase for a fixed velocity-space density is greatest (amounting to approximately 1 keV) at the highest observed energies, i.e. at a few keV. The largest relative increase (around a factor of 2.5) occurs at 20 eV, towards but not quite at the bottom of the observed range. All indications from the rather limited time series available are that the electrons in the wings are isotropic.

   If the wings are composed of accelerated solar-wind electrons, the acceleration is clearly highly energy dependent. It is highly reminiscent of the change occurring at the bow-shock ramp and in the ion releases, although exhibiting less tendency to form a plateau, and a greater tendency to change at higher energies.

6.2.1.3   Core

Within the 'core', we find a high degree of isotropy, and a distribution that is very similar to that of the wings at high energies and to the solar wind at low energies.

6.2.2   Interpretation

There are no firm 'theories' to test, established or otherwise, so we shall limit ourselves to some thoughts and suggestions.

6.2.2.1   Dilution by scattering

A first step is to look at the possible consequences of the magnetic field environment, as we did for the bow shock in Chapter 4. Figure 6.4 shows schematically that a tube of force, serving as a guide for electron motion contains two constrictions, one at each of the 'wings'. The consequences these constrictions would have for purely adiabatic motion (without acceleration and without scattering) of solar-wind electrons arriving from both directions are shown in the inset discs each of which is aligned with its corresponding field direction. Electrons of group **1** arriving with small pitch angles at **a** would reach

b with pitch angles extending, due to the increased field strength, over the full range of forward directions. They would proceed to **c** where, if the field had fallen to same value as at **a**, they would shrink into the original narrow range of pitch angles. At **d**, where for simplicity we take the field strength to be the same as at **b**, they would again occupy the full range of forward directions. On reaching the solar wind again at **e**—assuming the field strength not to have changed in the meantime—they would occupy once more the original angular range.

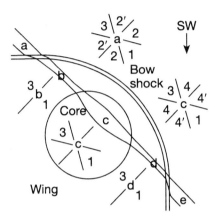

Figure 6.4. A possible configuration of a solar-wind event, indicating a bow shock and the various particle groups that may be involved. Display conventions as for figure 4.9.

Group **2** would be mirrored by the field gradient between **a** and **b** to return immediately into the solar wind as **2'** without reaching **b**. Groups **3** and **4** would behave the same as **1** and **2**, respectively, from the opposite direction.

The net result would be, as before, that the intensities and velocity-space densities at **a**, **b**, **d** and **e** would all be the same, the reduced area of the flux tube at the constrictions compensating exactly for the smaller numbers of electrons penetrating them. At **c**, however, the situation would be different. The electrons there would be observed to be counter-streaming at small (close to $0°$) and large (close to $180°$) pitch angles. There would be no electrons at pitch angles near $90°$ in this region. Another way of predicting the same result would be to note that region **c** with constrictions at both ends forms a 'magnetic bottle' within which particles at pitch angles near $90°$ would, if introduced, be trapped but where, if not, they would be absent. Electrons of groups **1** and **3** would not be trapped and could enter and leave without hindrance.

We observed, however, that the electrons at **c** were not counter-streaming but were approximately isotropic. If this isotropy was the result of the scattering of electrons from small pitch angles to cover the full range of angles, there would necessarily be a reduction of intensity, velocity-space density and number density. The reduction would be in direct proportion to the drop in field strength between **b** and **c**. This might be the effect which serves to keep electron velocity-space densities lower than typical magnetosheath values.

An 'explanation' of the acceleration in line with those of the bow-shock and release events is, then, that electrons are, depending on their energy, selectively accelerated in the 'wings' by the electrostatic waves found there, and that acceleration is accompanied by pitch-angle scattering. The scattering continues into the 'core', serving to 'dilute' the population by spreading it over the full range of pitch angles but without filling all possible trajectories.

6.2.2.2 Correction for dilution

If this is a valid explanation, it puts a slightly different complexion on the relative velocity-space densities of the 'wing' and 'core' seen in figure 6.3. The 'core' distribution resulting from acceleration of 'wing' electrons would, before scattering, have been approximately a factor of three higher than that observed. We have the temerity to make this 'adjustment' in figure 6.5 to reveal what might be the true, undiluted effect of acceleration, i.e. the distribution seen along an occupied trajectory. The 'core' distribution is seen in this light to cross the 'wing' distribution to give, as in all other examples of what are thought to be wave-produced accelerations, a greater proportion of high energies and a smaller proportion of low energies, thus constituting what may be described as 'heating'.

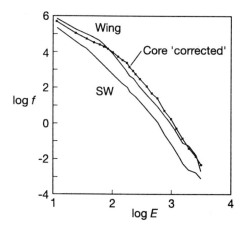

Figure 6.5. As figure 6.3, but with the core distribution 'corrected' according to an interpretation discussed in the text.

The 'core' distribution in other events, e.g. the one of 21 October 1984 depicted in figure 6.6, have been found to show this 'heating' relative to the 'wing' distribution even before 'correction' for dilution, so this does seem to be the true characteristic picture.

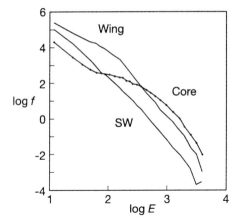

Figure 6.6. Electron energy distributions for the solar-wind event of 21 October 1984 [47].

6.2.2.3 Acceleration by lower-hybrid waves

The transition from the solar wind through the wings to the core is so strikingly reminiscent of that from the solar wind through the ramp to the magnetosheath and to the effects of the ion release (as comparison of the spectrograms will confirm) that the possibility of acceleration by waves applying here too must be considered.

The lower-hybrid resonance frequency ranges from ~ 7 Hz in the solar wind to ~ 20 Hz in the core of the 30 October event. Certainly waves were measured in the wings in this frequency range, though the coverage of the spectrum was not complete and the waves could not be positively identified.

Judging from the mean thermal energy of the ions, which rose steadily from ~ 10 eV in the solar wind to 350 eV in the core (A D Johnstone, private communication), resonant electrons would lie in this general range, rising in parallel with the ion thermal energy towards the core. This seems to be consistent with what is observed, at least to a rough approximation. Quantitative estimates of the wave strength required are beyond the scope of our present knowledge, primarily because the overall geometry and physical extent of the events is unknown. Future observations due to be made with several spacecraft simultaneously in the forthcoming Cluster mission offer the best chance of removing this major uncertainty.

6.3 Conclusions

Electron acceleration in 'solar-wind events' appears from its similarity to acceleration at the bow shock and in the ion releases to be explainable in terms of lower-hybrid wave acceleration of solar-wind electrons.

Chapter 7

Electron acceleration at the Earth's magnetopause

We now leave the solar wind and return to the Earth's magnetosphere. The magnetopause—the boundary layer separating solar and terrestrial plasmas—is an acceleration site with some distinctive properties which make it well worth our attention. This is a convenient opportunity to introduce, too, a new technique of analysis that considerably clarifies the nature of the transition and has potential value in other areas too.

7.1   The magnetopause

The transition between solar and terrestrial plasmas is not sharp but is effected through a boundary layer, or series of boundary layers within which transitions occur in all plasma and field properties. On some occasions the crossings are orderly as in plate 6(a); sometimes they are highly erratic as in plate 6(c).

The locations of magnetopause crossings recorded by UKS are shown in figure 7.1 [199]. As with the bow shock, the general locus is somewhat closer to the Earth than the average for this boundary owing to the extreme solar-wind conditions prevailing in late 1984 and early 1985.

7.2   A case study

The magnetopause crossing of 15 November 1984 depicted in plate 6(c) makes an ideal subject. Its position is ringed in figure 7.1. At the time of the crossing the magnetosheath field lay predominantly in the ecliptic plane. It was aligned with the average surface of the magnetosheath and pointed almost directly sunward. The magnetospheric field—in this case, that of the plasma sheet—was northward, as befits the equatorial region of the Earth's near-dipole. It was also slightly inclined, and lay parallel to the same surface as the magnetosheath field. The angle between the two field directions was 100°. The local conditions

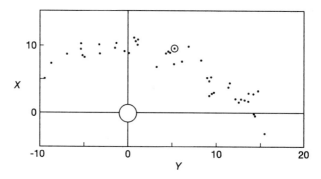

Figure 7.1. Magnetopause crossings by UKS. Coordinates are as for figure 4.2. The 'nose' of the magnetosphere was typically at a distance of 10 R_E, and the dawn flank at some 15 R_E. The crossing of 23 November 1984, used for the case study, is ringed (compilation by P A Smith).

were, then, somewhere between the two extremes depicted schematically earlier in figure 2.18.

7.2.1 'Perpendicular' electrons

7.2.1.1 *Energy distributions*

Figure 7.2 shows electron energy distributions in the magnetosheath, the plasma sheet and at a point in between. All distributions are for electrons travelling approximately perpendicular to the local magnetic field, with pitch angles of 80–100°. The magnetosheath distribution is, as would be expected, similar to those encountered in the bow-shock study, while that of the plasma sheet is the hottest yet discussed in these pages.

A noticeable feature is that all three curves intersect almost at a point. In fact, all distributions taken during this traversal of the magnetopause intersect at the same point. The energy where the common intersection occurs is 350 eV. We find this to be a feature of the overwhelming majority of events, the energy of common intersection ranging from event to event between 200 and 1000 eV.

This consistent orderliness simply begs to be made use of. Do we have to accept and try to make sense of the measurements in the chronological order in which they appear in plate 6(*c*) when there is clearly an underlying systematic progression from magnetosheath to magnetospheric conditions? We can try to exploit the underlying order to disentangle the rather chaotic picture and re-assemble the vertical columns of plate 6(*c*) in systematic rather than chronological order.

In order to be able to cope with the vast quantity of information contained in even a single energy–time spectrogram we shall need to contrive an automatic way of performing this task, but let us first establish the principle by hand.

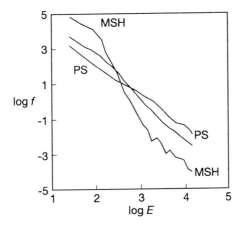

Figure 7.2. Energy distributions of 'perpendicular' electrons in the magnetosphere's plasma sheet **PS**, a point within the boundary layer (unlabelled) and in the magnetosheath **MSH**. Note that the three curves cross very nearly at a single point.

Plate 8(*a*) is a simplified version of plate 6(*c*). It is made up of just 24 vertical columns each representing an average over a short interval of time. The horizontal band of green at 350 eV corresponds to the common intersection point seen in the energy distributions. The same appears in the original spectrogram of plate 6(*c*), but it is harder to see there.

We now proceed with an exercise, which may be real or imagined, similar to that of solving a jigsaw puzzle. Make a photocopy of plate 8(*a*) (in colour if possible but this is not essential) and cut out the 24 vertical strips. Rearrange them to create the simplest graduated pattern of transformation from one extreme to another that you can. The result will, I suspect, be very similar to plate 8(*b*) (or its mirror image). In plate 8(*b*) the previously hidden gradual transition from the multi-coloured, narrow, cool spectrum of the magnetosheath to the broad, warm spectrum of the magnetosphere is now strikingly evident. It could be thought of as the solved jigsaw puzzle. In practice [199, 200] we perform this task automatically, and have found that all 30 UKS magnetopause crossings conform to this same general pattern.

As a bonus, we have discovered that other key quantities also revealed systematic patterns when put into the same revised order as the perpendicular electrons. Each orderly electron transition was accompanied by a gradual change in magnetic field strength (with a broad maximum reached partway across), a gradual veer of magnetic field direction, a steady reduction in the flow velocity of the ions and an accompanying increase in ion temperature. Before going on to examine the electrons travelling parallel to the magnetic field, the main issue of this chapter, we should just consider what the fact of this underlying pattern might imply.

The simplest interpretation of the underlying pattern is that the boundary layer between the magnetosheath and plasma sheet was moving back and forth, thus making the passage of the spacecraft across it highly erratic. If this is, in fact, the case, we can read the re-ordered spectrograms as representing re-constituted cross sections of the boundary layer. Taking note of Ockham's razor that we should not make things more complicated than they need to be (or, entities should not be increased unnecessarily [201]), this is the interpretation we shall adopt. In doing so, however, we need to bear in mind that matters just might be more complicated. For example, the very reason that causes the magnetopause to move—a change in conditions in the magnetosheath—almost certainly means that the samples taken in different parts of the medium will not apply to the same set of controlling conditions. Our interpretation of the cross section should not, therefore, be too rigid. In particular, we have no reason to expect the cross section as presented in a re-ordered spectrogram to be linearly related to distance across the layer. The pattern we have arrived at might well be stretched or compressed non-uniformly from the true spatial pattern. Such ambiguities are impossible to resolve from a single spacecraft, but should be well within the compass of a multiple-spacecraft mission such as Cluster. The order of the pieces should, however, remain inviolate, and we propose to exploit that small mercy now.

One way of interpreting the gradual transition between the magnetosheath and plasma sheet is to consider that there may be a mixing of the two families of electrons within the boundary layer. We should still keep in mind though the discussion of merging in Chapter 1, where we found that this process is possible only as an average effect of time and space variation of non-merging microscale trajectories. This mixing, as seen on the measurement scale, might result from a diffusion of high-energy electrons outwards from the magnetosphere and a diffusion of low-energy electrons in the opposite direction. If this were the case, the effects of both serve to heat the magnetosheath and to cool the magnetosphere.

According to this picture both sides would eventually attain the same intermediate distribution unless their characteristic distributions were maintained. If they were maintained, in the case of the magnetosheath by fresh electrons from the solar wind and in the case of the magnetosphere from a source maintaining the plasma sheet, a graduated boundary layer would be set up, as indeed the evidence strongly suggests. One of the most compelling items of additional evidence for the diffusion-like picture comes from the accompanying orderly reduction in ion flow velocity from, in this case study, 250 km s^{-1} in the magnetosheath to 20 km s^{-1} in the plasma sheet [199]. The boundary layer may be seen, then, as a region across which solar-wind momentum is imparted to magnetospheric ions through, in the absence of a true viscosity, the intermediary of plasma waves. From the discussion of distribution mixing in Chapter 1, it will be clear that the diffusion-like and viscosity-like processes are non-dynamical and non-reversible and are accompanied by an increase of entropy. While energy changes may, and are indeed likely to, accompany changes in position and direction, this interpretation would suggest that the energy of perpendicular electrons is, to all intents and purposes, conserved during inward or outward progression.

There are to date no models of how the apparent diffusion might be accomplished or of the profiles or graduations in velocity-space density at different energies to be expected. This is one of many areas calling out for attention.

7.2.2 'Parallel' electrons

Electrons travelling approximately parallel to the local magnetic field, with pitch angles of 0–20°, do not, as figure 7.3 shows, have a common crossing point. They can still be disentangled, though, using the key provided by the 'perpendicular' electrons. This is done in plate 8(*c*).

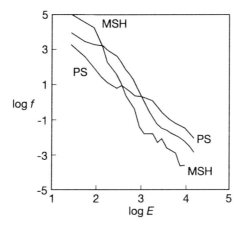

Figure 7.3. As figure 7.2, but for 'parallel' electrons. Here the traces do not have a common crossing point.

The parallel-electron spectrogram undergoes a simple transition only at the lowest energies. Medium and high energies have mid-boundary values that do not lie between the extrema. In this lies the first indication of acceleration at this site. There is in addition an ill-at-ease region where the high energies seem to belong deeper in the layer than the lower energies. There is similarity here with an earlier finding [202] where electrons in the outer part of the boundary layer displayed mid-boundary-layer velocity-space densities in one direction and magnetosheath velocity-space densities in the other.

Figures 7.4–7.8 compare energy distributions of parallel electrons with their perpendicular counterparts in the magnetosheath, the outer, mid- and inner parts of the boundary layer, and in the plasma sheet—respectively.

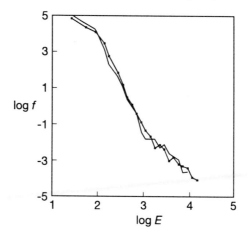

Figure 7.4. 'Parallel' electrons (smooth curve) and 'perpendicular' electrons (curve with points) in the magnetosheath.

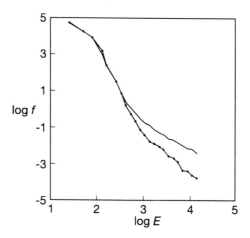

Figure 7.5. As figure 7.4, but in the outer part of the boundary layer.

In the magnetosheath (figure 7.4), the parallel electrons lie slightly below the perpendicular electrons, while the anti-parallel electrons (omitted for reasons of clarity) lie between the two, closer to the perpendicular ones. In the outer part of the boundary layer (figure 7.5), the parallel electrons match the perpendicular ones at low energies but have considerably higher velocity-space densities at high energies (this is the mis-match region of plate 8(c)). The anti-parallel electrons, on the other hand, are found to match the perpendicular electrons at all energies. Here, then, we have streaming in the magnetic-field direction, but not in the opposite direction. At the 'mid-way' point (figure 7.6) the parallel electrons and

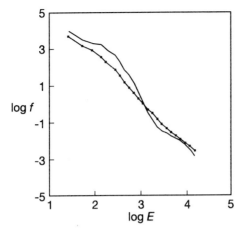

Figure 7.6. As figure 7.4, but mid-way through the boundary layer.

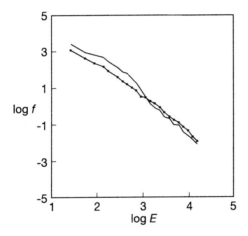

Figure 7.7. As figure 7.4, but for the inner part of the boundary layer.

the anti-parallel electrons 'bulge' away from their perpendicular counterparts at low and medium energies, representing counter-streaming at these energies. At high energies the electrons are approximately isotropic. At the boundary layer's inner edge (figure 7.7) counter-streaming continues at low and medium energies in the now considerably broader distribution. When the plasma sheet is reached (figure 7.8) the parallel and anti-parallel electrons match, but at low energies fall below the perpendicular electrons, which remain as at the inner edge of the boundary layer, to create a 'loss-cone' distribution typical of those adopted by trapped particles and fully understandable in these terms.

The counter-streaming 'bulges' at low energies in the inner and mid-

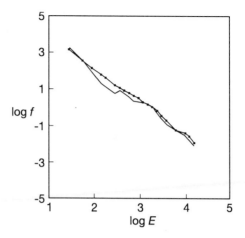

Figure 7.8. As figure 7.4, but for the plasma sheet.

boundary layer points strongly towards resonant field-aligned acceleration of a parent population similar to that corresponding to the local perpendicular electrons. If the lines of force threading these regions continued on to magnetic mirrors in both directions (as indeed is indicated by the loss-cone distributions at high energies) the accelerator could be either uni-directional (when counter-streaming would result from repeated mirroring) or bi-directional. A similar result would be produced on lines of force continuing to a magnetic mirror in one direction only if the accelerator were located on the remote side of the observing point to the mirror, but the loss-cone distribution at high energies argues against this alternative.

The striking similarity of inner and mid-region bulges with those at the bow shock and in the ion-release experiments strongly suggests a common cause—namely, acceleration by lower-hybrid waves. The magnetopause is, like all other regions so far studied, permeated by, amongst others, electrostatic waves, but the crucial frequency range is too poorly explored to reveal whether or not lower-hybrid waves are present in sufficient strength.

We have not, so far, come across an accelerator able to produce the parallel distribution found in the outer part of the boundary layer from the perpendicular one. Lower-hybrid acceleration could, in transporting electrons resonantly from low energies to high energies under favourable conditions generate the high-energy tail, but only at the expense of reducing velocity-space densities at the lower, source energies. It is perfectly clear in this case, though, that the curves do not cross, thus ruling out lower-hybrid acceleration as a cause. Adiabatic compression (however this might be achieved) fails to account for the difference. It would lead to an increase of velocity-space density by the same factor at all energies, which is clearly not the case here. A different explanation suggests itself. We noticed in the re-ordered spectrogram of plate 8(c) that for some three columns (representing many more energy scans in the original measurements) the high-energy electrons seem out of place. They match far more closely the high energy electrons found deeper within the region. This impression can readily be tested by tracing and matching.

If this correspondence can be accepted as a genuine clue, it suggests that, in addition to the scattering which leads to smooth transitions of the perpendicular electrons at all energies, high-energy electrons, preferentially, are scattered at low altitude from mid-boundary layer lines of force to outer-layer lines of force. The reduction in field strength between low altitude and high altitude would, then, reduce pitch angles on the outward journey to create the observed field alignment.

7.3 'Depletion layer'

A feature not evident in the above measurements, but which tends to occur over the dayside of the magnetopause in regions where there is little shear between magnetosheath and magnetospheric magnetic fields, is a slight (a few per cent) fall in ion temperature and mean electron energy as the density begins to fall from magnetosheath to magnetospheric values [203]. This localised drop runs counter to the behaviour during inward progression of our case study and in other crossings made by the UKS [199, 200]. Figure 7.9 compares electron energy distributions for perpendicular electrons in the magnetosheath and this 'depletion layer' obtained from the IRM at a time when the drop in mean energy was observed. What appears to have happened is that the higher-energy electrons of the magnetosheath have failed to penetrate into the outermost part of the boundary layer, whereas the lower-energy ones have done so—the net overall effect being a reduction in mean energy.

7.4 Flux-transfer events

The concept of a (magnetic) flux-transfer event envisages a temporary setting up of a magnetic-field topology such as that proposed for steady-state reconnection, as in the earlier figures 2.18(b) and 2.19. Such events are thought to occur primarily on the dayside magnetopause. Although the magnetic field will by definition change during a flux-transfer event, present treatments do not yet include the effects on the electrons of the electric fields bound to arise [204] from this. Plasma waves such as those referred to above might also be expected to be triggered by the reconfiguration and to have a rôle in particle acceleration. Some texts claim that the flux-transfer event is of overriding importance in magnetospheric dynamics [55, 205]. With this in mind we analysed all of the UKS crossings of the magnetopause (30 in all) to see if any had a significantly different behaviour to the normal gradual transition from magnetosheath to plasma-sheet properties reported on above. The result was that all 30 fitted

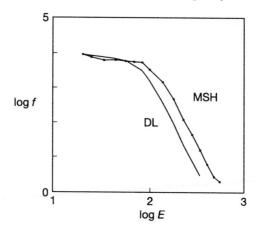

Figure 7.9. Perpendicular electrons in the magnetosheath **MSH** and in a depletion layer **DL** (from [203]).

in with the general picture, including one that had on other grounds been considered to be a flux-transfer event. It seems, then, that either all of the crossings happened to have been made during and relatively close to flux-transfer events or that none of them were. A recent suggestion [206] that all of the transitions were associated with flux-transfer events actually seems to show the opposite since a flux-transfer-event picture requires there to be a trough in magnetic field strength between the magnetosheath and magnetosphere, whereas the observations consistently show a ridge. It would seem, then, that there is no evidence in these data for the existence of flux-transfer events.

A useful next step in trying to establish whether the magnetopause does at times undergo changes that are qualitatively different to the norm would be to perform the analysis outlined above on very large numbers of crossings. If happenings of overwhelming significance do occur they would then surely make themselves known.

7.5   Conclusions

The magnetosheath appears to be a boundary layer through which 'perpendicular' electrons diffuse inward from the magnetosheath at low energies, and outward from the magnetosphere at high energies. Other key quantities also change systematically across the layer. This behaviour is consistent and appears, therefore, to leave little room for the invoked sporadic 'flux-transfer events'.

'Parallel' electrons undergo field-aligned acceleration within the layer. At least one magnetic mirror is involved in setting up the observed counter-streaming. Confirmation that in the mid- and inner parts of the boundary layer lower-hybrid waves are responsible for the energy-dependent acceleration of a population similar to that of local 'perpendicular' electrons awaits future measurements able to explore the energy range below 10 eV confirming the anticipated reduction in velocity-space density at low energies, and of course identification of the region's waves.

Magnetic-field alignment at high energies in the outer part of the boundary layer appears to be due to scattering from mid-boundary layer lines of force to outer-layer lines of force at low altitude.

Chapter 8

Electron acceleration in the Earth's magnetosphere

The Earth's magnetosphere is, as we saw in Chapter 2, the region of space dominated by the Earth's magnetic field and its contained plasma. The magnetosphere completely surrounds the Earth and forms its outermost environment. We saw how charged particles behave in its static or quasi-static electric and magnetic fields. We noted, though, that the magnetosphere is, in reality, a highly dynamic, constantly changing environment within which time-varying electric fields prevail and have the ability to accelerate charged particles. We do not yet have a reliable picture of the magnetosphere as a whole. Our grasp of the structure, content and changes has had to be assembled from isolated spot measurements from spacecraft pursuing a variety of orbits.

Following an account of the main distinguishable electron populations, and their possible origins, we shall examine each in turn and then discuss some of the theories that have been advanced to account for the powerful acceleration that is clearly taking place within this complex environment. These studies bring us into the realm of relativistic energies. We shall need therefore to work in terms of phase-space density rather than the velocity-space density that has, according to convention, been used up to this point.

8.1 Electron content

8.1.1 Sources

Figure 8.1 shows electron energy distributions of three of the main populations of the magnetosphere—the tail lobe, the plasma sheet and the outer radiation belt. The figure also notes typical distributions of the plasmas that border them—the magnetosheath on the outside and the ionospheric plasma of the plasmasphere underneath. Another source from underneath, the cosmic-ray albedo, is also shown. The figure includes, too, the distributions of cosmic-ray electrons that permeate the region, and solar and Jovian electrons that burst in from time

216

to time. In addition to these prospective sources are the electrons that appear throughout following the decay of cosmic-ray-albedo neutrons. These electrons are released into the medium with a flat distribution extending up to 782 keV [48]–[50] at rates falling off with distance from the atmosphere.

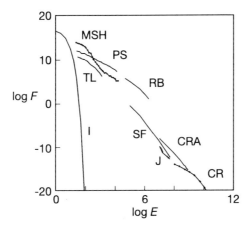

Figure 8.1. Electrons of the magnetosphere, with examples of the plasma sheet **PS** [47], the tail lobe **TL** [45], and the outer radiation zone **RB** [79] together with their possible sources—the magnetosheath **MSH** [47], the ionosphere, **I** (from [207]), solar-flare electrons, **SF** [82], electrons from the planet Jupiter, **J** (from [208]), cosmic rays, **CR** [80, 81] and cosmic ray albedo, **CRA** (from [209]).

It is clear at once that the electrons of the outer radiation belt have higher phase-space densities than other populations at the same energy, or, equivalently and more significantly, higher energies at a given phase-space density than their surrounding populations. Since phase-space densities cannot increase without acceleration (and even then can only do so under the special circumstances considered in Chapters 1 and 3) the outer-radiation-belt electrons clearly owe their existence to acceleration of surrounding plasmas sporting adequate phase-space density. These are the plasma sheet and ionosphere together with an admixture of neutron-decay electrons which build up with time. Whatever form it takes, the acceleration process appears to be one which imparts greater acceleration, relative and absolute, to higher-energy particles. The same applies to all but the lowest-energy component of the plasma sheet. The tail lobes have lower velocity-space density at a given energy than their neighbours the magnetosheath and the plasma sheet. On these grounds the tail lobes could, at least conceivably, be derived from them without acceleration by, for example, direction scattering. If, on the other hand, the lobes originate from the ionosphere, clearly a considerable degree of acceleration (several orders of magnitude at least) is required.

With these initial thoughts in mind, let us look at each of the key regions in turn, beginning with the tail lobes.

8.1.2   Tail lobes

Because of the very low electron intensities in the tail lobes, measurements are relatively poor. Detectors specifically designed to cope with this region would be inclined to saturate in the neighbouring regions, so a dedicated mission is really what is required. Such a mission, though, might not be seen to have the necessary 'glamour' to make it win funds in an intensely competitive field. For the moment, therefore, we have to make do with the rather sparse current information depicted earlier in figure 2.25 and reproduced here in terms of phase-space density in figure 8.2.

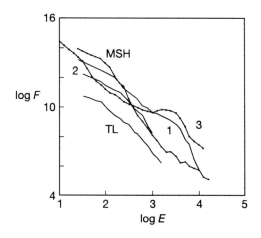

Figure 8.2. Polar-cap electrons **1** and **2** [45], and **3** [77] in relation to their possible sources: the magnetosheath MSH [47] and tail lobe TL [45].

Subjects of much greater attention than the tail lobes themselves have been the electrons precipitating from the lobes into the magnetically conjugate regions of the atmosphere centering on the north and south magnetic poles. Examples of these are shown in figure 8.2. The varying degrees of precipitation have been termed, according to their intensity and variability, 'polar rain' and 'polar squalls', as well as just 'polar-cap precipitation'. Unfortunately, we have no direct information on before-and-after distributions here since the tail lobes may be just as variable as the precipitation from them, so nothing can be assumed. One clear indication, though, that precipitation also involves acceleration is that the electrons are at times field-aligned. Acceleration could, as far as we know at present, be from magnetosheath energies, the source electrons finding their way in from this blanket of plasma or from the weak points of the northern and southern cusps in the magnetic field on the dayside of the magnetosphere. It appears, then, that the acceleration producing the higher energies is by at least an order of magnitude in energy, and may be as large as two orders of magnitude, depending on the state of the tail lobe plasma at the time.

8.1.3 Plasma sheet

The plasma sheet has been fairly thoroughly investigated. Electron distributions typical of the region have appeared throughout this discourse. They were discussed in relation to the aurora in Chapter 3, and compared with those of the magnetopause boundary layer in Chapter 7. Three of these and a further example are shown in figure 8.3. These alone establish that the plasma-sheet distribution is highly variable in magnitude and type. The shape of the region and the volume occupied are also highly variable, especially during magnetospheric substorms.

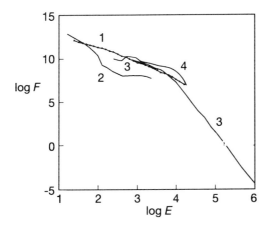

Figure 8.3. Examples of plasma-sheet electrons: **1** [47], **2** [46], **3** [69] and **4** inferred from precipitation at maximum phase of a pulsating aurora [210].

Some forms of precipitation from the plasma sheet—those producing the discrete forms of aurora close to the high latitude boundary of the region— were the subject of Chapter 3. Mentioned earlier, too, were the phenomena of glow and pulsating aurora which, as far as can be determined, take place at the conjugate, 'magnetic footprints' of the plasma sheet on the atmosphere in the auroral zones. It is generally understood that, no field alignment being involved, at pulsation maximum, the true intensity of the plasma sheet is experienced in the upper atmosphere as electrons are scattered so strongly that the loss cone is filled to the same density as the source region. Measurements taken even close to the atmosphere at pulsation maximum [210] may therefore be interpreted as samples of the plasma sheet itself. The pulsation-maximum trace in figure 8.3 illustrates further the variability of the region. Measurable intensities extend up to energies as high as 1 MeV.

8.1.4 Radiation belts

Radiation-belt electrons, i.e. electrons which because of their energies and the configuration of the magnetic field are (at least temporarily) able to execute the three adiabatic invariants of gyration, bounce and drift, occupy three main regions. These are the inner belt, which peaks at 1.4 to 1.8 R_E, the outer belt, whose maximum intensities are found from 4 to 5 R_E, and an intervening low-intensity 'slot' between 2 and 3 R_E [79]. The structure is not simple. As shown in figure 8.4, as we move inwards from the outer belt the higher energies show a fall to the slot followed by a rise, partially recovering the lost ground, to the inner zone. Lower energies, on the other hand, rise from the outer zone to the slot and remain at the same relatively high value into the inner zone.

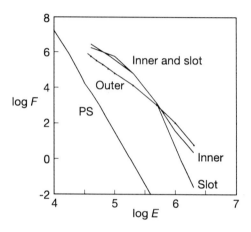

Figure 8.4. Electrons of the inner and outer radiation belts and in the intervening 'slot' [79, 208] shown in relation to a plasma-sheet distribution (**3** of figure 8.3).

Note that the distinction between radiation belt particles and plasma-sheet particles which share the same region of space, lies purely in their energy and, through their energy, their response to the fields permeating the region. Higher energies are more strongly affected by magnetic-field gradients while lower energies are more responsive to electric fields. The same applies, of course, and even more strongly to that other inhabitant of the innermost part of the radiation belt region—the plasmasphere. The electrons (and ions) of the latter are dominated, as we saw, by the co-rotation electric field. The plasmasphere holds the answer to the appearance of the slot between the two zones. The relatively high electron densities provide an ideal medium for the propagation of electromagnetic radiation such as whistler waves. These waves, generated by lightning discharges in the atmosphere and as man-made radio noise, scatter electrons into the loss cone from which, instead of bouncing back into the trapping region, they find themselves deposited into the upper atmosphere. The slot is thus a region of rapid erosion of high-energy electrons where densities remain comparatively low.

The outer zone frequently undergoes large (order of magnitude or greater) sudden (within minutes) increases of intensity, that have come to be known as 'injection events' [211]. These provide evidence of impulsive acceleration which requires explanation.

8.2   Acceleration theories

Theories to account for the complexity outlined above are still very much in their infancy. This is so much the case that there are some based, as you will no longer be surprised to learn, on conservative fields. We shall dispense with these first, then go on to consider attempts to form theories based on non-conservative processes, resonant and non-resonant.

8.2.1   Conservative fields

8.2.1.1   *Auroral potential-difference acceleration*

The much vaunted auroral potential-difference accelerator, which belongs in this category, was examined and, I trust, successfully exploded as a myth in Chapter 3.

8.2.1.2   *Steady-state reconnection*

As reported in Chapter 2, a process widely considered to take place within the magnetospheric tail as well as at the magnetopause is magnetic reconnection. Central to the process is a null in the magnetic field between opposing sources of field. In the magnetospheric tail the fields are those permeating the plasma sheet and the tail lobes. In the interpretation relevant to the current category the process is seen as occurring in steady state, with no essential time element [212].

The key features are usually depicted as in figure 8.5. This may be seen as an expanded version of the feature depicted in the magnetotail in the earlier figure 2.19. If one imagines the Earth beyond the left of the picture, the magnetic lines of force at the top continue on to the northern hemisphere of the Earth, while those at the bottom stem from the southern hemisphere. (The figure actually shows a selection of lines of force computed for the region between two attracting dipoles of the polarities indicated.) A localised region is envisaged where lines originally leaving the southern hemisphere for the deep tail return to the northern hemisphere via the loops at the left-hand side of the figure. Similar loops are at the same time created at the right-hand side. A dawn-to dusk cross-tail electric field—the same field as that discussed earlier in relation to magnetospheric convection—is considered to permeate the whole figure. Charged particles of either sign will $\mathcal{E}$-cross-B drift from the top and bottom towards the centre. On, or before reaching the central plane they will leave from the left and right under the same influences. The actual mechanism of acceleration is left unclear. It cannot, of course, be either the magnetic or the electric field, since both of these are static. Even temporary changes of energy are ruled out by the motion being entirely perpendicular to the electric field. Nevertheless, as I understand it, the suggestion is that plasma enters through large areas at the top and bottom and is expelled through smaller areas at the left and right at a higher speed in order to maintain continuity.

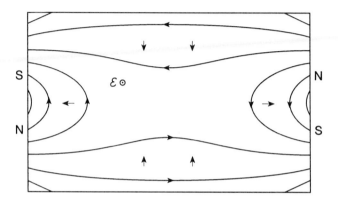

Figure 8.5. Magnetic null, central to the related concepts of (magnetic) re-connection and (magnetic) flux-transfer events. An electric field causes charged particles (of both signs) to $\mathcal{E}$-cross-B drift from the top and bottom towards the central magnetically neutral plane, and from the central null towards the left and right. The field topology shown here is actually that of the region between two attracting dipoles of the polarities indicated.

One further point should be noted. Magnetic topologies such as that of figure 8.5 are sometimes considered to fly apart to drive plasma islands, or 'plasmoids' [213] (see also [214]), tailwards. It is not immediately clear, though, how this is to be reconciled with the fact that the topology is that of two *attracting* magnetic elements.

8.2.1.3 *Serpentine acceleration*

The cross-tail electric field is also invoked in combination with the tail's magnetic neutral sheet to constitute a particle accelerator [215]. As summarised in figure 8.6, charged particles setting out to complete bounce motion from one side of the neutral sheet to the other are forced instead to execute a serpentine motion in the oppositely directed northern and southern tail magnetic fields. This motion is, as above, considered to carry electrons from lower to higher potential in the cross-tail electric field, thus accelerating them. Positively charged particles are carried in the opposite direction, and are also accelerated. This model, in common with many others, fails to take account of the counteracting effect of entry into the potential region.

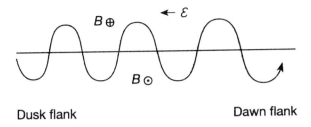

Dusk flank   **Dawn flank**

Figure 8.6. Serpentine motion of an electron gyrating between regions of oppositely directed magnetic field. One theory of acceleration holds that electrons (and ions snaking in the opposite direction) gain energy from a superimposed dawn–dusk electric field in the Earth's magnetotail.

8.2.2   Resonant accelerators

8.2.2.1   Landau resonance

Our wave model of auroral electron acceleration, based on Landau resonance between precipitating electrons and lower-hybrid wavepackets, appears to account well for the various properties observed in these electrons. The resonance requirement arises from the need for electron and wavepacket velocities to match, the tolerance on mis-match depending on wave amplitude. It remains to be investigated whether a similar process could operate to generate the various forms of polar-cap precipitation from tail-lobe plasma.

8.2.2.2   Gyro-resonance

Charged-particle magnetic moments can, as we saw in Chapter 1, be modified by electric fields oscillating at or near the gyro-frequency, in a continuous or quasi-continuous version of the principle used in cyclotrons in the laboratory. The rotating electric fields of electron-cyclotron and ion-cyclotron waves are invoked in some theories to account for acceleration perpendicular to the magnetic field within the magnetosphere. Acceleration perpendicular to the field automatically drives pitch angles towards 90°. Whistler waves, whose frequencies occupy a range immediately below the electron gyro-frequency, may also resonate when their frequency is Doppler-shifted upwards as seen by the approaching electrons [216]. This latter phenomenon is invoked primarily as a modulator of the precipitation rate rather than as an accelerator.

8.2.2.3 Drift resonance

Resonance with electric fields rotating at or near particle drift periods is thought to be one of the main causes of the acceleration of magnetospheric particles up to radiation belt energies [217]. Figure 8.7 shows the combined effects of inward drift and acceleration produced by an electric field rotating in resonance with gradient drift. While the magnetic flux contained by the drift orbit (the third invariant) is reduced, the magnetic moment (the first invariant) remains constant. Energy will thus increase in proportion with magnetic field strength. Particles whose drift phase is 180° out of resonance will be driven outward and retarded. Under the influence of plasma waves having suitably slowly rotating electric vectors, charged particles from any source tend to diffuse throughout the region of the magnetosphere where the magnetic field is orderly enough to support such drift motion. This populates inner positions with energetic particles and outer positions with low-energy ones. Factors of energy increase would, in the simplest version of the model, be the same at all energies. This phenomenon could account, at least qualitatively, for the transition shown in the lower-energy component of figure 8.4 from the outer radiation zone to the inner zone, which amounts in this instance to something approximating a uniform factor of three. The rise at high energies from the slot to the inner zone does not fit such a simple picture since there the factor of energy increase at constant phase-space density increases with energy.

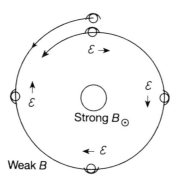

Figure 8.7. Energisation of a trapped electron by a rotating electric field. The magnetic field, directed out of the paper, falls off in strength with increasing distance from the Earth (central circle). The guiding centre of an electron gyrating in this field gradient and in a rotating azimuthal electric field $\mathcal{E}$ progresses inward and gains energy as in the earlier figure 2.17. If the electric field local to the electron rotates in phase with the electron's drift, as shown, the electron will become steadily energised as it progresses inwards. If the field is out of phase the electron moves outward with diminishing energy.

When trapped or temporarily trapped particles become energised during inward migration, their combined gyratory and bounce motion occupies smaller and smaller tubes of force. As a consequence the northern and southern mirror points become closer together. This is sometimes seen as being equivalent to the closing of mirror points in Fermi's second cosmic-ray accelerator (see Chapter 10). The decreasing distance is then considered to be the agent through which the particle becomes accelerated as though it were being compressed by a piston. However, although the particle on its inward journey certainly experiences in its frame of reference a shortening of the distance between mirror points, there is in fact no change in the magnetic field in the observing frame of reference in which acceleration occurs. These particles are not, therefore, accelerated by closing mirror points any more than a ball is accelerated by being bounced between walls set at an angle. If the particles are scattered in pitch angle as they migrate inwards, they will tend to occupy the whole tube of force and may appear to be energised adiabatically by being compressed into a smaller volume. Equally, this is an apparent rather than a true compression. The analogy does afford, though, a guide to the degree of acceleration that might result when we note that the volume of a tube of force varies as the fourth power of its linear dimensions. A gradual inward drift from the 4 R_E centre of the outer zone to the slot region at $2R_E$, might, therefore be accompanied by—and in fact be dependent on—an acceleration of as much as 2 raised to the power of 4×1.67, which at more than 100 is more than adequate to account for the observations.

8.2.3   Non-resonant accelerators

The dynamic and fluctuating nature of the magnetosphere gives rise to changing magnetic fields and thereby to induced electric fields. Possibilities for acceleration are constrained to some extent by the fact that magnetic fields cannot indefinitely increase or decrease in magnitude. There has to be a recovery phase.

8.2.3.1   *Induction*

One of the possibilities that has been raised is that electric fields induced from the constant re-structuring taking place in the magnetosphere may have a component parallel to the magnetic field and be stable enough over periods of tens of minutes to an hour or so to account for auroral electron acceleration [218, 219]. There is a need for a model of this mechanism to establish whether it could account for the many well established features that are observed. Would the process lead to conjugate aurorae, for example, or would the two hemispheres be in anti-phase? Should we expect to see upward-streaming electrons and downward-streaming field-aligned ions when the field changed sign? Would the fields be strong enough and extensive enough to accelerate magnetospheric electrons to energies of 30 keV or more? These are at present unanswered questions.

We discussed in Chapter 2 how a charged particle drifts in longitude under the influence of a magnetic field with a gradient perpendicular to B. What happens if the field strength changes with time? If the changes are slow compared with the drift period the particle preserves the total magnetic flux through the drift orbit, the third adiabatic invariant. The radius of the drift orbit increases for non-relativistic particles in proportion to the particle's velocity, and for ultra-relativistic particles in proportion to the kinetic energy. The particle's guiding centre moves to keep the magnetic field strength at the guiding centre constant.

The magnetic fields of the magnetosphere caused by and at the same time controlling particle motion in the ring current are subject to variation resulting from compression (or expansion) of the magnetosphere in response to changes in the solar wind, and to modulation by Alfvén waves and other modes. These changing fields will produce betatron acceleration or retardation, with electrons and ions being affected simultaneously and in the same sense. Acceleration that increases magnetic moments, as in this case, will drive pitch angles towards 90°, while retardation will lead to field alignment. I am not aware of any models by which the full range of possible effects of betatron acceleration have be explored. If the third adiabatic invariant is conserved while slow, large-scale changes take place to the magnetic field, acceleration will be accompanied by outward displacement, and retardation by inward migration.

A numerical study of the possible effects of the steadily increasing northward-directed magnetic field that occurs in the neutral sheet within the plasma sheet during substorms has discovered that although the motion is sometimes very complicated favoured particles can gain energy steadily from excursions into the region of induced electric field [220].

8.2.3.2 *Magnetic annihilation*

It has long been realised that the adjacent regions of oppositely directed magnetic fields that develop in the solar corona might undergo mutual annihilation, so releasing the energy previously stored in the magnetic fields. Some of this energy could power particle acceleration [221, 222]. The process differs crucially from the steady-state reconnection considered earlier in that in the case of annihilation the before and after states of the magnetic field are different. While the overall principle that magnetic-field energy will be freed during such events is indisputable, the means by which this energy—stored originally in the charged particles carrying the currents which give rise to the fields—is conveyed to the particles accelerated by the annihilation, remains unclear.

8.2.3.3 *Magnetosonic waves*

One explanation [223] for the electron and proton injection events has been prompted by the fact that an event of this kind has recently been observed to follow a sudden compression of the magnetosphere, caused by enhanced solar wind reaching the magnetosphere following a solar flare. The compression was modelled as a magnetosonic wave which propagated from the magnetopause to the ionosphere and back again. The wave was represented for the numerical simulation by a pulse of azimuthal electric field with associated magnetic perturbations. The rises and falls of intensity at a notional satellite, duly progressing along its orbit, as the particles drifted around the magnetosphere, have been well reproduced with suitably selected initial conditions. The results suggested that the electrons were driven in from the outer radiation belt, while the protons were a mixture of solar-flare protons that had penetrated to the outer belt and had subsequently been driven inward, and inner-belt protons accelerated locally. The first adiabatic invariant was conserved in the modelling by both species of particle. Energies of the non-relativistic protons duly increased in proportion to magnetic field strength, while electron energies, being in the relativistic range, duly increased with only the square root of the field strength.

8.3   Conclusions

This brief survey of the electrons of the Earth's magnetosphere shows that the process of acceleration is instrumental in forming the many and varied populations from their sources, the solar wind (via the magnetosheath), the ionosphere, and the products of cosmic radiation. Acceleration is instrumental too in some of the loss processes, such as those leading to the discrete aurora and some forms of polar-cap precipitation. There are many possible contributing processes. A common factor here is a time-varying electric field, in the form of a wave or as the induction resulting from a varying magnetic field. Some of the processes currently in vogue appear to be purely conservative and therefore non-viable.

Chapter 9

Electron acceleration at the Sun

The Sun is a particle accelerator *par excellence*. Not only does it generate the solar wind which is the driving force for particle acceleration in and around the Earth's and other planetary magnetospheres, but it produces in the explosive events known as solar flares [224, 225] copious numbers of electrons, protons and heavier particles up to the ultra-relativistic energies which classify the Sun as a cosmic-ray source, albeit a relatively minor one. There are seemingly insuperable difficulties, though, in trying to investigate its workings. Due to the hostile environment of the Sun's outer atmosphere, or corona, where much of the acceleration is thought to take place, there are, as yet, no direct measurements of either sources or immediate products. Neither are there measurements of the electric and magnetic fields which create one from the other. Enough is known, though, from indirect measurements made from the Earth's environment and from deep-space probes to make a tantalising and intriguing puzzle.

9.1 Regions of the Sun [224]

9.1.1 Interior

The Sun is a relatively cool star. Its core is a nuclear fusion reactor, building helium from hydrogen. The released energy is transmitted outwards through the core, an intermediate interior and a convection zone through a sequence of processes including radiation, absorption, re-radiation, diffusion and turbulent convection, in the course of which the temperature falls from its core value of some 8×10^6 to 6.6×10^3 K. The internal workings of the Sun represent a rich area of study.

9.1.2 Photosphere

The first layer visible to us is the photosphere. This is located at a helio-centric radius of $\sim 7 \times 10^5$ km. The photosphere is only some 100 km thick and being so sharply defined is considered to form the 'surface' of the Sun. Certainly most of the solar mass of 1.99×10^{30} kg is contained below it. The temperature of the photosphere, as derived from the widths of emission lines, is around $\sim 6.6 \times 10^3$ K. The temperature at first continues to fall as height is gained, falling to 4.3×10^3 K at an altitude of about 600 km.

9.1.3 Chromosphere

The temperature begins to rise again through the thick (several thousand km) tenuous layer above the photosphere, known as the chromosphere. This heating is understood to be derived from absorption of upwelling material and sound waves. At the top of the chromosphere—some 200 km above its base—the temperature rises to over 2.5×10^4 K.

9.1.4 Transition region

Between 2000 and 2500 km the temperature begins a steep climb through a transition region to the base of the outermost region—the corona.

9.1.5 Corona

The temperature in the corona may reach as high as 2 MK. The cause of this final step-like rise remains one of the many mysteries of the Sun. The corona extends into space, as the solar wind, out to the heliopause where its sovereignty finally ends. The corona is highly inhomogeneous. It displays radially elongated bright areas, or streamers and dark, slightly (30%) lower-temperature, and much lower- (order-of-magnitude) density areas, known as coronal holes.

On occasions there is a greatly enhanced, localised emission of coronal material in events known as coronal mass ejections, CMEs [225]. These represent when they reach the Earth, after some four days of outward travel, gross disturbances in the size and even topology of the Earth's magnetosphere and, through these, the ionosphere. A recent exceptionally large event was witnessed all the way from the Sun to its impact on the Earth's outer environment [226]. Such disturbances promote some of the acceleration processes discussed in the previous chapter.

9.1.6 Solar flares

A solar flare [224, 225] is an explosive, transient, localised disturbance in the corona. Through the enhanced emission of radio waves, visible light, x-rays, gamma rays and energetic particles it is the means of release of 10^{22} to 10^{25} J of energy previously accumulated in coronal magnetic fields. These magnetic fields take the form of gigantic loops or arches extending out from the Sun. Flares last from minutes to an hour or more. The mean rate of power dissipation varies from 3×10^{20} to 10^{24} W (more energy in each second than mankind has used in all history [224]). Some of the charged particles accelerated in a flare are guided by the magnetic field down into the chromosphere and some into the relatively dense photosphere below. Here they generate a wide range of electromagnetic emissions in secondary processes. Other flare particles escape into interplanetary space. Characteristic products of the accelerated electrons are x-rays generated by synchrotron emission in the remaining magnetic fields.

9.2 Solar-wind electrons

The properties of the solar wind are well known from many spacecraft and deep-space probes that have explored the heliosphere from a 0.3 AU outwards. Measurements taken close to the Earth at 1 AU, just outside the bow shock, naturally form the bulk of these. The vast quantity of data that have ensued tend, however, to be more tantalizing than revealing about the origin of the electron component of the solar wind. The reason is fundamental. It is that due to the great (orders of magnitude) reduction in magnetic field strength between the point of delivery—the corona, if indeed this general supposition is correct—and the point of observation, very few of the observed electrons could have travelled adiabatically from the Sun. Those doing so would be confined to very small pitch angles. The collimation would be an extreme version of that experienced by auroral electron albedo or the emergence of electrons from the magnetosheath into the solar wind. The majority of those observed would have needed to have been scattered very substantially by, say, magnetic field structure on the scale of a gyro-radius or by resonant plasma waves. Coulomb collisions would be too infrequent to play an important role. Since the scattering process is likely to be energy dependent, the distribution actually observed may be very different to that leaving the Sun. This 'through a glass darkly' situation will remain until a spacecraft built to withstand the severe environment of the corona is sent to the Sun. Such missions have been proposed, for example [227], but at present we must make the best of what we have.

9.2.1 Observations

9.2.1.1 Energy distribution

We saw earlier that electron distributions encountered by the UKS tended to a first approximation to follow a negative power law in velocity-space density (and, at these non-relativistic energies, in phase-space density) of $\gamma \approx 3$ over the measured range (10–800 eV), and could be slightly better represented by kappa distributions. However these told only part of the story. Comprehensive studies of solar-wind electrons performed by the Heos spacecraft which penetrated in to 0.3 AU, and by others reveal many more intriguing features [228]. In particular, the electron distribution is frequently either noticeably or even strongly elongated parallel or anti-parallel to the magnetic field in the sense to give a net flow away from the Sun. Elongation sometimes appears in the towards-the-Sun direction also—or instead of the parallel direction.

Figure 9.1 shows a set of contours for a relatively broadly elongated distribution. Figures 9.2(*a*) and (*b*) show energy distributions parallel and perpendicular to *B* for a pronounced away-from-Sun elongation, figures 9.3(*a*) and (*b*) the same for a broad elongation, and figures 9.4(*a*) and (*b*) the same for an isotropic distribution. The (*a*) parts of these figures are in log *f*–linear *E* format, where Maxwellians would appear as straight lines: the (*b*) parts are in log *f*–log *E* format where power laws would be straight lines.

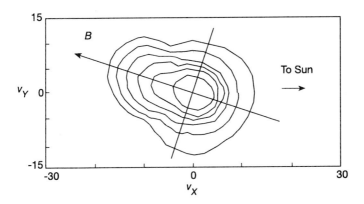

Figure 9.1. Solar-wind electrons flowing preferentially in the magnetic field direction in the sense to take them away from the Sun. Velocities range from zero to ±30 Mm s⁻¹ in V_X (radially from the Sun in the ecliptic plane), and ±15 Mm s⁻¹ in V_Y (towards dawn). Contours of velocity-space density are spaced at one order of magnitude, the highest nearest the centre (from [228]).

One school of analysis (see [228] and references therein) seeks to describe the distributions as composites formed by the merging of two Maxwellians—a low-temperature 'core', and a high-temperature 'tail'. It is immediately clear, though, that none of the distributions in figures 9.2(*a*), 9.3(*a*) and 9.4(*a*) are well described this way, as indeed the experimenters themselves conclude from quantitative goodness-of-fit tests. The UKS measurements seen earlier fare no better, as figure 9.5 confirms.

9.2.2 Acceleration theories

There are at present no theories to account for the acceleration of solar-wind electrons. Discussion to date has been concentrated on trying to understand how an electron population assumed to reside at the base of the solar corona might escape into interplanetary space.

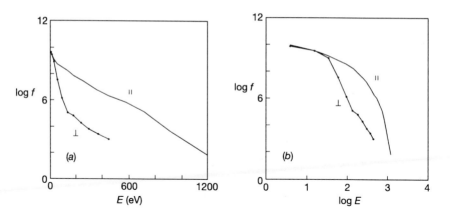

Figure 9.2. (*a*) Electron energy distributions parallel and perpendicular to the magnetic field for a pronounced away-from-Sun elongation as in figure 9.1 (from [228]). (*b*) As (*a*), but with a logarithmic energy scale.

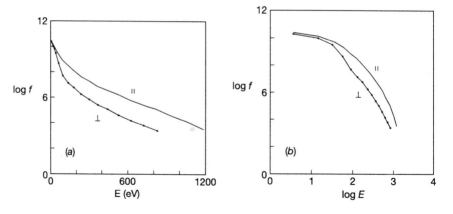

Figure 9.3. (*a*) Electron energy distributions parallel and perpendicular to the magnetic field for a broad elongation (from [228]). (*b*) As (*a*), but with a logarithmic energy scale.

9.2.2.1 Velocity filter

Escape is considered to be impeded by Coulomb collisions and by an assumed negative potential of the interplanetary medium with respect to the Sun [229]. A key aspect of this idea is that the higher-energy electrons have only the potential barrier to climb, while the lower-energy ones have to contend also with collisions with the ambient medium because cross sections for collisions are greater at lower energies. The corona is therefore seen as a energy or velocity

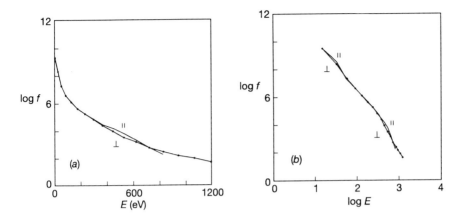

Figure 9.4. (*a*) Electron energy distributions parallel and perpendicular to the magnetic field on an occasion when isotropy prevailed (from [228]). (*b*) As (*a*), but with a logarithmic energy scale.

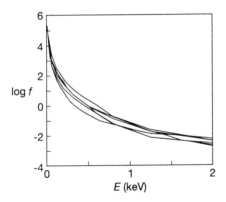

Figure 9.5. Electron energy distributions observed by UKS, seen earlier in figure 2.2, but now with energy on a linear scale.

filter.

This model envisages two electron components within the corona—high-energy electrons about to overcome the potential barrier and therefore potentially 'free', and low-energy electrons unable to do so and therefore hopelessly 'trapped'. The escaping high-energy electrons, having been retarded by these processes, then occupy a narrow range of pitch angles in the weak magnetic field of interplanetary space to give the elongated features shown above. The remaining volume of velocity space receives its electrons by scattering from this allowed cone of directions. Wave–particle interactions are also seen as possible contributors. Distributions found in interplanetary space could become further shaped by changes in electrostatic potential within the medium resulting from the interplanetary electric field.

In this picture then, the elongated high-energy tails are thought to reflect— after attenuation within the corona and possibly after subsequent modulation— the electron population generated by an unknown mechanism at the base of the corona. The perpendicular component arises, in this model, as a by-product, and carries little information about coronal electrons.

9.2.2.2   A statistical model

Let us consider another possibility. We cannot fail to be struck by the similarity between the contour plot of figure 9.1 and the equivalent representation of the auroral electron stream seen in figure 3.10. There, we sought, with some success, to understand the auroral distribution as being created by a magnetic-field-aligned acceleration of an originally isotropic distribution. It is also noticeable that the transitions between the perpendicular and parallel distributions in figures 9.2 to 9.4 have much in common with the transitions from the solar wind to accelerated solar wind at several of the previously examined sites of acceleration, i.e. the bow shock, the ion-release disturbances and the natural events occurring in the solar wind. Could a similar explanation apply? It would certainly be in the spirit of Ockham's razor [201] to seek a common cause for a common effect.

Following this new line of interpretation, if the elongations are generated in interplanetary space by an energy-dependent, or resonant, acceleration at localised shock fronts for example, the perpendicular distribution rather than the parallel distribution would be the one more closely related, however distantly, to that released from the Sun. If so—and far more investigation is needed to test the hypothesis—we are looking for a source, an acceleration mechanism and an escape process leading to a distribution most simply described as a near power law whose exponent increases steadily with energy.

We found in Chapter 1 that 'dynamic scattering' leads to a near-power-law distribution whose slope increases from one value at non-relativistic energies to another at ultra-relativistic energies. However this simple picture leads at low energies to

$$F \text{ or } f \propto E^{-1}$$

i.e: a negative exponent of only 1, much less than the observed values of around 3.

Power laws of steeper slope can, however, be produced at non-relativistic speeds if each interaction is a multiple one composed of several scatterings. To investigate, let us consider scattering centres that move towards one another in some cases, and move apart in others. Let us consider that each step of energy gain or energy loss consists of not one but a sequence of reflections between approaching or receding reflectors so formed. Let each gain or loss be limited by the time for which the closing or opening pairs of mirrors are maintained. In this case the number of velocity reflections experienced will depend on particle velocity. Fast particles will have more reflections than slow particles, and the slowest may experience none at all. A computer simulation of this simple scheme shows that the energy steps will be proportional to energy, the constant of proportionality having a value for non-relativistic particles that is double that for ultra-relativistic ones. A power law is therefore possible in principle by these means if suitable scattering centres exist in the corona. At present this is no more than conjecture.

It has to be admitted that we are a long way from understanding how solar-wind electrons are generated.

9.3 Solar-flare electrons

Electrons of considerably higher energy than those normally found in the solar wind are emitted in bursts during solar flares. The nearest we can come to observing the acceleration process is to monitor, from a spacecraft that is clear of the Earth's magnetosphere and associated perturbations, the burst of particles which commonly appears within minutes of a flare being observed. The burst reaches a maximum an hour or so later and then gradually decays away over many hours.

9.3.1 A case study

9.3.1.1 Observations

Figure 9.6 [82] gives an example. Here we see the time profiles of electrons in four different energy bands in terms of the count rates of detectors borne on two interplanetary monitoring platforms IMP-6 and IMP-7 following the solar flare of 7 September 1973. The flare, located at latitude and longitude S18 W46, reached maximum brightness at 1212 UT. Allowing for the Sun-to-Earth transit time of the light bringing this information, the maximum brightness at the Sun would have been 500 seconds, or about 8 minutes earlier at 1204 UT. Elapsed time in figure 9.6 is measured from this base. The flare is recorded as beginning half an hour earlier than this, and ending 1.5 h later, so it was a comparatively long-lived one. We see that the electrons in all energy bands produced a maximum count rate after an elapsed time of about 2 h. They all then gradually decayed away over 1.5–2 h.

9.3.1.2 Arrival times

What can we learn or infer about the electron distribution released from the Sun from such sequences? The first point to note from figure 9.6 is that the flood of electrons began to be measurable—i.e. clearly distinguishable above the pre-flare background—about 20 min after 'flare maximum' at the Sun. Now, the fastest electrons in the lowest energy band (i.e. those at 38 keV) would have

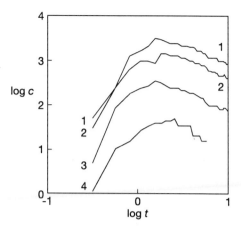

Figure 9.6. Electrons arriving at the Earth's orbit from the solar flare of 7 September 1973, the subject of the case study. Count rates are for: **1**, electrons of 20–38 keV; **2**, 38–76 keV; and **3**, 75–342 keV measured from IMP-6; and **4**, electrons of 200 keV–1 MeV measured from IMP-7. There is a four-decade scale of count rate. The abscissa is the logarithm of time in hours from the arrival at the near-Earth satellites of light from the flare (from [82]).

taken at least 23 min to reach the Earth (if they were not speeded up or slowed down on the way), so these must have left the Sun before or soon after flare maximum. The maximum count rate at the highest energies (200 keV–1 MeV) occurred about 2.5 h after flare maximum. For these particles the rectilinear transit time is about 10 min, so, even if they had left the Sun as the flare waned, there is still a full hour to account for during which these particles would have meandered approximately six times the Sun–Earth distance. Electrons of all energies were still arriving at the Earth many hours after the flare had finished, again indicating great distances covered along very tortuous paths. These facts all point towards the electrons not travelling freely through interplanetary space but being gradually diffused by being scattered from irregularities, in the interplanetary magnetic field.

9.3.1.3   *Random walk in interplanetary space*

Let us imagine that the electrons were suddenly released from the Sun into interplanetary space at flare maximum, say, and let us suppose that interplanetary space contains magnetic irregularities able to scatter them. Further, let us endow these scattering centres with the ability to scatter any electrons encountering them in random directions, some continuing forward, some returning sunward and others (most) with a combination of radial and azimuthal motion. For the moment we consider the scattering centres to be stationary so that there is no change of energy when scattering takes place. The average distance an electron travels between scatterings, determined by the size of the scattering centres and their spacing, is known as the mean free path, λ.

At first, and at distances of the order of a mean free path from the Sun the particles will emanate from the source region in what might be seen as a tidal wave. Some of those observed will arrive at such a closely positioned observer promptly, without any scattering, while some might be delayed a little by having undergone deflection. Close to the source the highest fluxes and highest densities will occur almost at once. Further away, many mean free paths away, there will be few prompt arrivals. Most particles will have suffered many deflections, with the consequence that the highest intensities will occur after elapsed times of several rectilinear travel times. The times of maximum intensity in figure 9.6 indicate that there are very few prompt arrivals. It seems clear, then, that a considerable degree of scattering is involved in reaching the Earth's orbit.

The early stages of such a process are very difficult to assess analytically. However, for later times an analytical approach known as diffusion theory—a method of analysis which follows the consequences of systematically increasing entropy—shows that following instantaneous release of N_E particles per unit energy from a point source into an unbounded medium, the number density per unit energy at distance r is given by

$$n_E(r, t) = \frac{(N_E/4)\exp\{-3r^2/4v\lambda t\}}{(\pi v\lambda t/3)^{1.5}}$$

or, when expressed in terms of differential intensity,

$$j(r, t) = \left(\frac{N_E c}{16\pi}\right)\frac{\sqrt{E^2 + 2EE_0}}{(E + E_0)}\frac{\exp\{-3r^2/4v\lambda t\}}{(\pi v\lambda t/3)^{1.5}}.$$

The distributions of n_E and j in radial distance have the same form as the distribution in velocity under the influence of random increases and decreases in energy, i.e. a Maxwellian, the rms distance being $\sqrt{4/3v\lambda t}$. The reason is that the same governing principle applies; it is that the distribution seeks to maximise entropy, i.e. to attain the most disordered, flattest state permitted by the constraints. Since there are no constraints here equivalent to the restriction on total energy which create a Maxwellian, the distribution can continue to flatten indefinitely. It differs from an ink spot spreading on blotting paper where the ink particles eventually come to rest. These distributions do not; they continue to spread.

The development anticipated by the formula above is shown in figure 9.7. Since the total number is conserved, the number density at the source ($r = 0$) falls steadily with time, once the diffusion stage has been reached. The rate of spread and the rate of fall are naturally governed by the particle's speed (faster particles spreading more quickly) and the mean free path (longer mean free paths extending the spreading further). The overall rate of progress is governed by the product of these two factors.

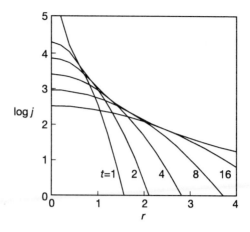

Figure 9.7. Radial diffusion profiles. Intensity as a function of radial distance at selected times following release from a point source at $t = 0$. The envelope shows how the maximum intensity diminishes with increasing radial distance.

To see whether the observed time profiles actually conform to this formula we exploit the fact that the equation for j can in the best tradition be moulded into a straight line by taking the logarithm of both sides and re-arranging to give

$$\log(jt^{3/2}) = A + B/t$$

where A and B depend on N_E, v, λ and r.

This states that, under the conditions laid down above, the logarithm of the intensity multiplied by the elapsed time raised to the power 3/2 will give a straight line when plotted against the reciprocal of elapsed time. Moreover, the slope of the line is

$$B = -3r^2/4v\lambda$$

from which, r and v being known, λ can be found. The result of this test for our four electron components is shown in figure 9.8. They all fit quite well [82]. We shall not lose our main line of discussion here by considering just how good the fit is, or whether the count rates would fit equivalent expressions for one-dimensional or two-dimensional diffusion, except to say that the test is, in fact, not a particularly sensitive one. Nevertheless, the count rates do conform to our very simple hypothesis. The mean free paths resulting from figure 9.8 are different at the different energies, ranging in a more detailed analysis covering all available energies, from 0.24 AU at 20 keV to 0.1 AU at 10 MeV. These relatively long mean free paths serve to confirm our suspicions gained from the early 'times to maximum' that we are, at 1 AU, witnessing an early stage of diffusion to which the formula does not really apply. The different mean free paths for the different energies are also disquieting aspects of the analysis, especially when the lower-energy particles seem to have been deflected less than their higher-energy and consequently higher-rigidity fellows.

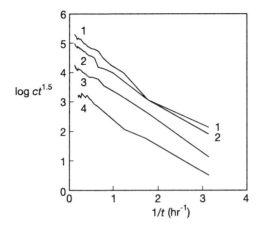

Figure 9.8. Testing for radial diffusion. Analysis of the counts rates of figure 9.6, according to a method discussed in the text. Simple diffusion through interplanetary space would yield straight lines.

This analysis technique just described was invented as part of an exploration of solar-flare protons [230] recorded by the satellite Explorer 12. It stemmed from noticing that log–log plots of intensity against, not time, but time multiplied by velocity, i.e. distance travelled, were the same shape at all energies. This immediately revealed that all protons, regardless of their energy, were going through the same motions, the only difference being that they were doing so at different rates. This test is, of course, independent of the actual shape of the time profile and is consequently more general. Since relative intensities between different proton energies were the same at all stages of development (measured in terms of distance travelled), the relative intensities were also those that must have obtained at the point of release, i.e. at the solar flare. The relative intensities could then be taken as a measure of the energy spectrum produced by the solar flare itself. Diffusion analysis gave us the extra information that the mean free path was typically 0.04 AU for all proton energies.

9.3.1.4 Source spectrum

Before leaving the simple-diffusion model, and in lieu of anything better, it is interesting to pursue the idea further in case that subsequent results, in this arena or elsewhere are found to be consistent with a simple-diffusion interpretation. By doing so we see how the initial spectrum obtaining at the source could in principle be derived from the measurements at a remote location.

The intensity reaches a maximum at a time $t_{max} = r^2/2v\lambda$. Substituting

t_{max} into the expression for j, we find that

$$j_{max} \propto \frac{1}{r^3}.$$

Since this is also true for a point close to the source (as long as the requirements for diffusion are met), and j_{max} at the source is by definition the source intensity, the spectrum of j_{max} observed at radial distance r is directly proportional to, and provides a useful measure of the shape of, the source spectrum [230].

A distribution composed from the maximum intensities at each energy is shown in figure 9.9. The distribution exhibits the typical features of a near-power law at lower energies and a higher power law at higher energies, with a transition at around 100–200 keV. This is the best estimate that can be made at present of the electron distribution produced by acceleration in the vicinity of solar flares.

In view of the inconsistencies we found above between the results of the analysis and the assumptions initially made, we have to be highly sceptical about whether or not this estimate, which is quite typical of those found in other flares, can be trusted and used to test theories of acceleration. Some help is at hand here from measurements of the x-rays and gamma rays emitted during solar flares. Estimates can be made of the quantities of these radiations expected from the best estimate of the source distribution for direct comparison with the radiation measurements. Such tests turn out to be very reassuring and so give some confidence that the best estimate distribution is at least a reasonable approximation to the truth. Consistency does not amount to confirmation, but it is a step on the way. Much thought and endeavour has been invested in attempting to improve the model of simple diffusion to try to take into account the many additional factors that might be expected to play a role in the real world [231]. These include acknowledgment of the fact that the magnetic field of interplanetary space has the garden-hose pattern discussed earlier causing particles to diffuse more readily along this direction than across it.

Large-scale emission of enhanced plasma from solar-flare regions may also produce large-scale coherent structures in interplanetary space which make the diffusion non-uniform. Very importantly, the scattering centres are almost certainly in motion. Scattering centres moving outward with the underlying solar wind will tend to be moving apart as they take up an increasingly larger volume. Scattering from such centres will lead to systematic loss of energy, by adiabatic cooling. In a few cases [231] it has been possible to compare electron intensities at two distances from the Sun, the near-Earth position at 1 AU and a spacecraft at 6 AU. Such comparisons have given some idea of the way the mean free path increases with radial distance. These sorts of study are really very difficult and have to be tackled numerically (with the consequent loss of generality), but are thought wholly worthwhile since it is only by such methods that we can investigate the capabilities of the Sun as a particle accelerator. Proposals that are currently mooted for spacecraft to travel close to the Sun [227] would clearly yield invaluable results pertinent to this key question. The aim is to get as close as 4 R_S or 0.02 AU, i.e. within the average mean free path deduced above. This would clearly represent a huge advance over the previous closest approach by Heos of 0.3 AU, just inside the orbit of Mercury.

9.3.2   Acceleration theories

9.3.2.1   Particle sources

For the second time in these pages we encounter an acceleration problem in which we have no direct knowledge of the electron distribution before acceleration. We cannot even hope to obtain such knowledge because the environment of the flare is so hostile that no spacecraft could survive the temperature and radiation.

We are left, therefore, with no option but to attempt to reason what the initial distribution might be. The natural assumption is that it is similar to that of the solar wind. Coronal temperatures, and the solar wind reaching the Earth suggest, then, that the initial energies lie predominantly in the 1–100 eV band.

9.3.2.2   Processes

There are few suggestions as to the processes involved in transferring energy from motion and magnetic fields into charged particles.   General suggestions include dynamic scattering and 'compression' between magnetic irregularities rushing outwards with the expanding plasma, and irregularities in the surrounding medium. There has recently been some discussion over whether a flare's magnetically stored energy finds its way directly into the electrons or whether it is transmitted first to protons and other ions, and subsequently on to electrons [232]. None of these is at present far enough advanced to help in our present quest. We shall, however, make some suggestions of our own with regard to dynamic scattering.

9.3.2.3   Dynamic scattering

One very striking fact, and possibly important clue to what is going on, is that at ultra-relativistic energies, say above 3×10^5 eV, the distribution takes the form of a power law of negative exponent $\approx$ 4.5, but at non-relativistic energies the slope is much reduced. We noted in Chapter 1 that for sub-relativistic energies the distribution resulting from dynamic scattering would indeed be of this type, the slope at very low energies gradually reducing to give a negative power law of exponent 1, whatever the magnitude of the exponent at relativistic energies. As a test of whether the observed distribution might be accounted for by such a process figure 9.9 shows a modelled distribution found by trial and error to give the best fit to the observations. This requires a value of

$$N_0\beta^2 = 0.06.$$

The fit achievable cannot be described as good. The penalty for a close fit at high energies is too rapid a falling off at lower energies. A better compromise between magnitude and slope could be found, but this would conceal the significant disparity. However, any euphoria over a good fit would need to be tempered by the knowledge that the derived source distribution is subject to major uncertainties, and the model, derived with a steady state of affairs in mind, is unlikely to apply too closely to an explosive temporal event such as a solar flare. Nevertheless, we do have, on the face of it, a reasonable fit—there is a change of slope, and this occurs in the right place, i.e. in the region of transition from non-relativistic to ultra-relativistic energies.

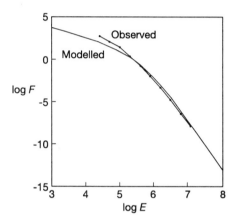

Figure 9.9. Comparison of the electron energy distribution evaluated for the case study solar flare of 7 September 1973 [82] with a distribution predicted for acceleration by 'dynamic scattering' in the flare region.

If we set aside the above reservations, what can we learn from the best fit value of $N_0\beta^2$? Even without separating the two terms we can begin to gain some insight into the workings of this type of accelerator by considering how much energy is likely to be gained by a particle leaving the system after making the expected, or typical number N_0 of gain-or-loss interactions.

The typical factor of energy gain, i.e. the excess of gains over losses, is

$$G_0 = (1 + \beta)^{\sqrt{N_0}}.$$

If N_0 is large

$$G_0 = \exp\{\sqrt{N_0\beta^2}\}.$$

On substituting the result above we find that the typical gain is only 1.3.

To explore further, we need to need to break down the product $N_0\beta^2$. If the scattering and reflection is produced by moving patterns of magnetic field, the velocity of these centres, and hence β, will be determined by the Alfvén velocity v_A, which is in turn determined by the magnetic field strength B and the mass density ρ of the medium, i.e.

$$\beta \sim \frac{1}{1 + (c/v_A)^2}$$

where

$$v_A = B/\sqrt{\mu_0 \rho}.$$

With an expected field strength of 0.04 T and a mass density of 10^{-11} kg m^{-3}, $\beta \sim 0.04$. N_0 then becomes 43. If N_0 is determined by escape from the medium by random walk through its boundaries, we can estimate the mean free path λ from the approximation

$$N_0 \sim (R/\lambda)^2$$

where R is the scale size of the medium. If we make $R \sim 10^6$ m, we get $\lambda \sim 10^5$ m.

We see now that the time taken to complete the typical number of interactions, $(N_0\lambda/v)$, is only 0.02 s, if $v \sim c$. This is easily accommodated within the typical flare duration. Of crucial interest, however, is the time it would take for electrons remaining within the region to have their energy raised from solar-wind energies of ~ 10 eV to the observed 10 MeV, i.e. by some six orders of magnitude. This may be estimated by noting that the typical number of interactions for this, N, can be found from

$$10^6 = (1 + \beta)^{\sqrt{N}}$$

which gives $N \sim 10^5$, requiring a time of less than 100 s—still well within the observed flare duration.

There seem, therefore, to be no obvious barriers to this interpretation. There is no positive indication, though, that the process actually operates.

9.3.2.4 Wave turbulence

A recent suggestion [233] on different lines calls attention to the fact that turbulent space plasmas appear almost universally to give rise to electrons and (more particularly here) ions with 'ring' distributions in velocity. Such distributions are unstable to the generation of lower-hybrid waves which in turn are readily consumed by electrons, as we have discussed in other chapters. Numerical methods were employed to see how an electron distribution might evolve in response to a hypothetical initial ion ring distribution. The distribution evolves, not surprisingly, and as predicted for the other areas, becoming steadily flatter and extending increasingly further towards higher energy. The results of one of the studies have been found to fit quite well the deduced source distribution at lower energies, but the high-energy bounds have not yet been explored due to prohibitively long computing times.

9.4   Conclusions

There are currently no theories able to account for the acceleration of solar-wind or solar-flare electrons, major causes of uncertainty being lack of knowledge of the initial distributions, the coronal environment and modulation during interplanetary propagation. In the case of solar-wind electrons, escape from a potential well may have a part to play in shaping the distribution that emerges from the Sun, as may dynamical scattering. Dynamical scattering may also be involved in producing solar-flare electrons, as indeed may wave turbulence.

Chapter 10

Electron acceleration in the Cosmos

The greatest acceleration mystery of all remains that of the acceleration of cosmic rays. Cosmic rays are charged particles—predominantly protons, but with an admixture of heavier nuclei, electrons and positrons—which pervade, it is believed, the whole Cosmos [49, 234]. They enter the solar system from sources and acceleration regions located throughout the galaxy and very probably beyond. In the case of the protons and heavier nuclei, energies range from the lowest capable of penetrating the magnetic field of the heliosphere i.e. $\sim 10^6$ eV, to the highest energies known to be attained by individual nuclei, i.e. 10^{20} eV— and possibly even beyond this [235]. Our study of cosmic-ray electrons, or perhaps better just cosmic electrons, which though very energetic do not reach these extreme energies, will nevertheless raise many of the key issues related to the origin of the nuclei as well. It may as well be admitted at the outset that the problem is quite baffling.

10.1   Electron observations

10.1.1   Direct

Figure 10.1 shows an energy distribution of cosmic electrons drawn from measurements made with ionisation chambers carried on high altitude balloons. At the lowest energies there are two curves—the lower applying to times of high solar activity, when the magnetic barrier of the heliosphere was correspondingly enhanced, and the higher to times when the Sun was relatively quiet. The modulation effected by solar activity and consequently the state of the magnetic fields in the heliosphere testify to the difficulty electrons of energies below 10^{10} eV have in penetrating these heliospheric fields.

Between 10^{10} and 3×10^{11} eV the energy distribution takes on what has become the familiar form of a power law. In terms of phase-space density, the negative exponent is close to 5. This corresponds, in this ultra-relativistic regime, to a negative exponent of 3 in intensity, the usually quoted form.

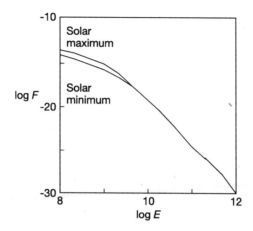

Figure 10.1. Cosmic-ray electrons at periods of maximum and minimum solar activity (from [80] and [208]).

At the highest measurable energies, i.e. the highest energies where adequate numbers of electrons can be detected, the distribution falls more rapidly than in the near-power law applying in the middle range of energies [236]. It is here that present measurements come to an end. Because of the low and increasingly rapidly falling fluxes of electrons, further exploration becomes harder by ever-increasing degrees. A similar situation prevails at the top end of the nucleon distribution, around nine orders of magnitude higher. It is one of the many tantalising aspects of cosmic rays that the energies that might be the most telling about the whole mechanism of cosmic-ray production remain just—only just—out of reach.

Successful attempts have been made in recent years to detect the positively charged anti-matter counterpart of the electron—the positron—in cosmic radiation [224]. These have demonstrated that positrons are outnumbered more than ten to one by electrons. This result is of great importance since it shows at once that the electrons do not arise from the decay into electron–positron pairs of π^0 mesons, produced in nuclear interactions suffered by cosmic-ray nuclei, since this otherwise promising candidate would result in equal numbers of electrons and positrons.

10.1.2 Indirect

High-energy electrons reveal their presence in remote locations through the electromagnetic radiation they emit when gyrating in magnetic fields [237]. Gyration involves acceleration (in the general sense), and consequently results in the emission of radiation. This magnetically induced radiation is known as synchrotron radiation from the radiation from the same cause which hinders acceleration in these laboratory devices. The radiation is equivalent to that arising from acceleration (again in the general sense) from electrostatic attraction by atomic nuclei, known as bremsstrahlung (braking radiation). For this reason synchrotron radiation is sometimes referred to as magnetic bremsstrahlung.

The rate of loss of energy through radiation by a relativistic electron (or positron) of energy E gyrating in a magnetic field of B nT is [238]

$$dE/dt = 3.9 \times 10^{-22}(EB)^2 \text{ eV s}^{-1}.$$

The frequency of the emitted radiation covers a broad band with the most probable photon energy

$$\hat{E}_{\text{photon}} \sim 7 \times 10^{-24} E^2 B \text{ eV}.$$

Figure 10.2 shows the type of electromagnetic radiation corresponding to $\hat{E}_{\text{photon}}$ as a function of E and B. These relations show that an electron at the top of the observed range, say 10^{12} eV, travelling through the interstellar magnetic field of ~ 0.1 nT, radiates in the infra-red. The distance it is able to travel before being seriously degraded is $\sim 10^{19}$ m, equivalent to the thickness of our galaxy, the Milky Way. This might not seem much of a restriction (!), but in cosmic-ray terms it represents a significant limitation. It means that, however the sources of cosmic electrons are distributed throughout interstellar and intergalactic space, the observed energy distribution is likely to be severely attenuated at and above 10^{12} eV [239]. The fact that the measured distribution shows a falling-off from the power law in just this range of energies is consistent with this expectation, and at the same time suggests that the $\gamma = 5$ power law may be the true primordial product of the accelerator.

Electrons of less than the extreme energies considered in the above example will, in the same fields, radiate at lower frequencies. This radiation is in fact the main source of a continuum of radio noise observed from all directions in space [224]. The radio emission is patchy and indicative of a far-from-uniform distribution of magnetic fields. In the relatively strong fields of supernova remnants and active galaxies, emission can extend into the optical, x-ray and, possibly, gamma-ray bands. In these denser media, electrons can be expected also to lose energy by radiating x-rays via bremsstrahlung.

10.2   Acceleration theories

If we accept the strong inference that the electron energy distribution below 10^{10} eV is determined and controlled by the heliospheric magnetic fields, and that the distribution at the very highest energies is degraded by synchrotron losses, we are left with a power law of negative exponent 5 as our best measure of the 'local' cosmic electron distribution. The situation is therefore similar to that found in solar-flare and other electrons except that the exponent here is now somewhat higher.

What are we to make of the power-law distribution?

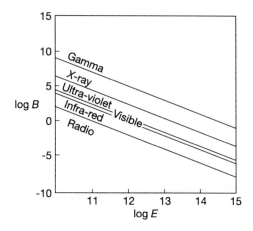

Figure 10.2. Synchrotron radiation emitted by electrons of energy E (in eV) gyrating in magnetic fields of strength B (in nT).

10.2.1 Fermi's first theory

We begin by paying homage to the famous theory by Fermi [2] to which many present-day acceleration ideas owe their origin and for which the highly schematic figure 10.3 sets the scene. Fermi considered that cosmic rays (only the nucleonic component was known at the time) might be accelerated, above an initial injection energy, by elastic interactions with 'wandering magnetic fields' in the galaxy. Interactions would take place via the trajectory-bending effect of magnetic fields embedded within gas or plasma clouds moving through space. The interactions might be of the type in which a particle is deflected as in a reflection by a moving mirror, as in figure 10.4(a), or by a moving magnetic irregularity as in figure 10.4(b). He pointed out that, whatever the actual process, a charged particle would gain or lose energy in momentum transfer with the gas cloud. In the simplest case of reflection back along the incident path, i.e. the strongest interaction, an ultra-relativistic particle's energy would change by a factor of $1 \pm 2\beta$ (to first order in β) where β is the velocity of the moving mirror v relative to that of light. In recognition of the fact that many of the interactions would be glancing and therefore less effective, $1 \pm \beta$ was seen as the typical factor of change. The positive sign applies for head-on collisions, while the negative sign applies for overtaking collisions. Fermi pointed out that head-on collisions were statistically more likely than overtaking ones (for the same reason that car windscreens get wetter than rear windows), the probabilities ψ being in the ratio of the relative velocities of the charged particle and approaching and receding gas clouds i.e.

$$\psi_{\text{gain}}/\psi_{\text{loss}} = (c + v)/(c - v).$$

This formula, much quoted in cosmic-ray texts today, has I confess, caused me some worries. The term $c + v$, seems to imply relative velocities greater than c. The relative velocity of two objects moving in the same direction at u and v is not $u - v$ but

$$v_{rel} = (u - v)/(1 - uv/c^2).$$

If either u or v, or both u and $v = c$, then $v_{rel} = c$.

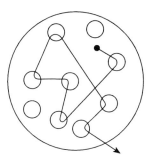

Figure 10.3. Schematic representation of the elements of Fermi's first cosmic-ray acceleration theory. Particles emitted from a supernova or other source 'collide' at random with moving regions of enhanced or otherwise non-uniform magnetic field before escaping from, or being lost within, the region.

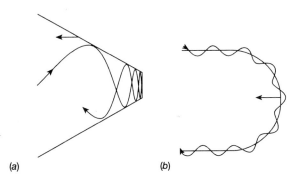

(a)   (b)

Figure 10.4. (*a*) Acceleration of a cosmic ray by reflection from a moving magnetic mirror: the 'type-A' interaction of Fermi's first theory of cosmic-ray acceleration (understood from [2]). (*b*) Acceleration of a cosmic ray by spiralling around a curved line of force which is in motion: the 'type-B' interaction of Fermi's first theory of cosmic-ray acceleration (from [2]).

However, any confusion is resolved very easily when it is realised that the factor applying to the question under discussion is not the relative velocity (the velocity of one body in the reference frame of the other) but simply the difference in the velocities as seen from a frame in which the velocities are u and

v. The ratio of the gain and loss probabilities as expressed above is therefore perfectly described by the above ratio.

As a matter of interest, it is of course perfectly possible to evaluate the ratio of gains and losses from any reference frame including that of one of the objects as long as due account is taken of the fact that in the strange, non-intuitive world of relativity, the spacing of approaching and receding gas clouds is different when seen from a fast-moving particle because of the relativistic distortion of distances (I am indebted to G I Powell for discussions on this point).

Fermi pointed out that the statistical bias favouring gain would lead to a particle steadily gaining energy. He showed that the average rate of gain would, as we have seen in Chapter 1, be proportional to energy such that

$$dE/dN \sim E\beta^2.$$

Energy thus increases exponentially with the number N of scatterings or steps, i.e.

$$E = E_1 \exp(N\beta^2)$$

E_1 being the initial energy. A given consignment of particles, e.g. from a stellar flare or a supernova, would then, in the absence of other factors, steadily rise in energy as N increased with time. To prevent this continuing indefinitely Fermi introduced another factor, that of a loss of particles due to the hazards of collisions with interstellar material and/or to loss from the acceleration region. These would curtail N and introduce a mean number of scatterings N_0 which particles might or might not exceed by chance. The losses are, in this model, compensated by a steady rate of injection from the sources so that at any given time there would be a mixture of young weakly energised particles and older more energetic particles (a mixture bucking the usual human trend!). The probability of surviving for between N and $N + dN$ scatterings when the mean is N_0 is

$$\psi(N, N_0) = (1/N_0) \exp\{-N/N_0\}dN.$$

By noting that

$$\psi(E, N_0) = \psi(N, N_0)dN/dE$$

the equivalent expression for attaining energy between E and $E + dE$ is found through replacing N by $\beta^2 \ln(E/E_1)$ from above to get

$$\psi(E, N_0) = (E/E_1)^{-(1+2/N_0\beta^2)}.$$

The number density and (since all of these particles have speed c) the intensity are thus power laws of negative exponent for this systematic process

$$(\gamma_S)_{\text{intensity}} = 1 + 2/N_0\beta^2.$$

The phase-space density is also a power law and its exponent is

$$\gamma_S = 3 + 2/N_0\beta^2.$$

This is one of the curves in figure 10.5. The observed phase-space density exponent of approximately 5 for nucleons (it is the same, too, for electrons) suggested that $N_0\beta^2$ would need to be approximately 0.5. There was, and still is, no information on what the two components of this product should be separately. Note that for much larger values of the product, γ_S reduces to what might be seen as the ultimate attainable value of 3. At this limiting value, the number density of high-energy particles would become infinite. For a negative exponent of 4, the energy content of high-energy particles would become infinite, making this higher value the practical lower limit for the exponent, i.e. for the flatness of the distribution attainable by these means.

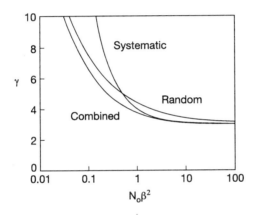

Figure 10.5. Exponents of negative-power-law energy distribution predicted for purely **Random** gains and losses by random collisions with moving scattering centres, for **Systematic** energy gain, and for the two effects **Combined**. The exponents γ are shown as functions of the product of the average number of collisions made before escape or loss from the system N_0 and the square of β the mean velocity of the scattering centres relative to the velocity of light. Note that all processes lead ultimately to the same exponent of 3.

It was a remarkable achievement to predict the form of the cosmic-ray distribution—which, it will be noticed, is a composite one—from fundamental considerations. There were, however, some unsatisfactory features noted by Fermi. One of these was the curious fact that the product of two apparently independent quantities, N_0 and β^2, turned out to be the same at all energies and that its magnitude was of the order of unity. To this day, no physical reasoning has been given for either the magnitude or the energy independence. Another was the extreme slowness of a process which depends, as this does, on second-order effects, i.e. on β^2 when β is necessarily a very small quantity.

Yet another, that has entered in more recent decades with the added knowledge that the nucleon distribution extends up to energies exceeding 10^{20} eV, is that magnetic fields in the interstellar medium are, on average anyway, too weak to deflect such high-energy particles within distances of the order of the size of the galaxy itself. To imagine that the deflection centres are of the sizes of galaxies and that the galaxies are themselves vast moving magnetic mirrors is somewhat precluded by the huge path lengths that would be required and the inevitable collision losses and losses by and to radiation that would result even if a time equal to the age of the universe, i.e. the Hubble time estimated at approximately 10^{10} years [224] could be contemplated for the process.

10.2.2 Fermi's second theory

These limitations prompted Fermi to put forward a more efficient process [240] for the acceleration of cosmic-ray nuclei. Instead of the more-or-less cancelling gains and losses that would ensue from randomly moving scattering centres, the new process envisaged some particles gaining energy systematically by being caught between closing magnetic mirrors. The mirrors were thought to arise from modulation of the large-scale galactic magnetic field by magnetohydrodynamic waves. While this second mechanism does overcome the slowness of the first mechanism, its somewhat contrived nature makes it somewhat implausible. It still carries the disadvantage of requiring much stronger or much more extensive magnetic fields than are currently believed to exist in the galaxy or in intergalactic space. It would though be unwise to discard the mechanism on this last ground alone because so far no mechanism that has been proposed has found a way round this crucial problem.

There seems to be no reason why such a mechanism might not play a part in the lesser task of the acceleration of cosmic electrons.

10.2.3 Random walk in energy

A factor not taken into account in the analysis of either of the two mechanisms discussed above is the broadening of the energy distribution that results from random gains and losses in energy, an effect discussed in Chapter 1 and applied to solar-flare electrons in the last chapter. Only a few years after Fermi's first theory was advanced, Davis [17] combined the notions of entropy increase with Fermi's wandering magnetic fields and found that a negative power law is again the result at relativistic energies. The phase-space density exponent for this random process is

$$\gamma_R = 3 + \sqrt{2/N_0\beta^2}$$

a result that can also be obtained using a statistical approach [18]. This is compared with γ_S in figure 10.5.

For a negative exponent of 5 we require

$$(N_0\beta^2)_R = 0.5.$$

The result is fortuitously exactly the same as that resulting from the earlier analysis of Fermi's first mechanism.

The result of the statistical process may be seen in terms of a tendency towards the unrestrained high entropy $\gamma = 3$ composite distribution, curtailed by loss or escape. The degree with which the distribution falls short of the ultimate exponent of 3 is ameliorated by greater magnitudes of β and N_0. Whether this process would be enough in practice to account for the electrons is subject to exactly the same considerations as the Fermi first mechanism.

10.2.3.1 *Random walk and systematic energy gain combined*

It is natural now to consider what happens when the two effects are combined. Davis examined this situation also [17]. He found that

$$\gamma_{R+S} = 2 + \sqrt{1 + 2/(N_0\beta^2)}.$$

Again this reduces to the ultimate exponent of 3 for large $N_0\beta^2$. Figure 10.5 includes this function, too, for comparison. For the observed exponent of 5 we require now

$$(N_0\beta^2)_{R+S} = 0.125.$$

These extra considerations have therefore made little difference to the requirements for the product which seems to haunt these studies.

10.2.3.2 *Implications*

What can we learn from the fact that, when both systematic and random effects are allowed, the best-fit value of $N_0\beta^2$ is 0.125? Firstly, we can note, using the reasoning given in the previous chapter, that the energy gained by a typical particle is $\exp(N_0\beta^2)$, i.e. only 1.42. In order to proceed further we have to know N_0 and β^2 separately. To enable us to explore one of a very wide range of possibilities, let us consider the moving magnetic mirrors to be products of Alfvén-wave turbulence in interstellar space. Interstellar space is a highly diverse medium, with a wide range of ion mass densities and magnetic-field strengths. For present purposes we shall adopt nominal, conservative (low) values for both, of 10^{-21} kg m^{-3} and 5×10^{-10} T. These combine to give an Alfvén-wave velocity of 1.4×10^4 m s^{-1}. So we have for illustration

$$\beta \approx 5 \times 10^{-5}$$

making

$$N_0 \approx 5 \times 10^8.$$

We can gain some feel for the distance scales involved for this model to work by noting that the gyro-radius of the highest-energy electrons the process appears to be capable of producing, 10^{13} eV, is, in a perturbation field five times stronger than the typical, $\sim 10^{13}$ m. The mean free path would, in this picture, need to be an order of magnitude larger, at 10^{14} m. The linear scale size of the acceleration medium would be $\sqrt{N_0}$ times greater still at $\sim 5 \times 10^{17}$ m. This is comparable with the thickness of the galactic disc, and so is certainly large but not completely out of the question.

So far so good. But, for this picture to be tenable we still need to estimate how long it would take to accelerate electrons by the required amount, and to ensure that this can be accomplished in a time substantially less than the age of the galaxy. Using the same procedure as in Chapter 9, we find that the time required for acceleration from the output of solar and presumably stellar flares, say 10^6 eV to 10^{13} eV, i.e. a factor of 10^7, is $\sim 10^9$ yr. As it stands, this is very tight. Even if we juggle with the parameters of the model, we find that time required is really quite robust, and really constitutes something of a problem for the model. The reason can be appreciated at once by noting that it takes 4×10^5 yr for a typical particle to gain its factor of 1.4 in energy, whereas the model envisages high flyers avoiding escape or loss long enough to build up by a very much larger factor. It was the inherent slowness of this type of random process that Fermi found unsatisfactory. It remains one of the fundamental tantalising cosmic-ray problems to which no answer is yet forthcoming.

10.3   Conclusions

The low-energy portion of the cosmic-ray electron distribution in the vicinity of the Earth may be understood as being governed largely by heliospheric magnetic fields presenting a barrier through which the particles have to diffuse from their origin beyond the heliosphere. The barrier is harder to penetrate when the Sun is more active.

The medium-energy portion of the distribution which appears to reveal the true state of cosmic electrons is of the negative-power-law form to be expected from random energy gains and losses that are proportional to energy, combined with losses and escape from the acceleration region. The product of the mean number of scatterings and the square of the fractional energy gain seems to be between 0.1 and 0.5. The reason for this magnitude, and the reason it is independent of energy, remain total mysteries—a cosmic conundrum indeed. The rapid falling off of the distribution at the highest energies is entirely understandable in terms of energy losses due to synchrotron radiation in transit. In fact Fermi understood the absence of (then measurable) electrons in just this light.

Chapter 11

Reflections

We have explored ten sites of electron acceleration in the natural plasmas of space. Energies and energy changes have ranged from the lowest currently measurable to the highest theoretically attainable. We have striven throughout to understand what lies behind this acceleration—what are the mechanisms at work, and how are they powered?

Specific problems have been posed by a wealth of knowledge gained from measurements from many types of space vehicle by numerous experimenters. Guiding lights for interpretation have been the classical dynamics of Newton, the continuity constraints summarised by Liouville, and the statistical mechanics of Maxwell and Boltzmann. Particular springboards have been Fermi's and Davis's ideas on acceleration by random processes, and Landau's concept of wave–particle interactions. We have heeded too the keep-it-simple philosophy of William of Ockham.

It has been necessary to extend existing theories to meet the challenges posed by the space plasmas. The generation of peaks from previously monotonic distributions has forced us to step outside the realm where Liouville's theorem of continuity holds, i.e. into non-dynamical, irreversible, dissipative systems. These peaks have also forced us to extend the ideas of Landau from those which lead to plateau formation. Fermi's and Davis's cosmic-ray acceleration concepts have also played a large part in our thinking, and these too have been extended from the original ultra-relativistic range to include, for solar-flare electrons, the non-relativistic regime as well. Our wave model of the aurora has been modified in turn to suit the different conditions pertaining at the bow shock and other dynamic areas in the solar wind.

What we can be certain of—as certain as we are that energy is conserved—is that acceleration to higher speeds involves, in space as it does in the laboratory, temporal change. This vital element may be provided by a changing electric or magnetic field, or by an unchanging but moving configuration of either. Continuous acceleration naturally demands replenishment of particles and energy.

Several widely held interpretations have had to be discarded when it was discovered that they contained only static electric and magnetic fields and were, therefore, inert.

It will have been apparent that, despite our best efforts, we are still some way from being able to resolve these questions with any degree of certainty. Perhaps the closest to being understood is auroral electron acceleration. We can be fairly confident in picturing the acceleration as taking place just an Earth radius or so out along auroral-zone magnetic lines of force lying in or near the boundary between the plasma sheet and the tail lobes of the magnetosphere. The unstable nature of the relatively sharp transition between these plasmas gives rise to waves—lower-hybrid waves seem best to fit the bill—which draw their energy from plasma sheet ions and which are rapidly consumed by electrons. The stop-press news is that the waves now actually appear to have been observed in the acceleration region. However, we still need to know whether wave amplitudes and velocities are really capable of delivering the observed degree of acceleration. Another missing item is a good description of the particles and fields of the tail lobes. This is crucial since it appears to be the difference between tail-lobe and plasma-sheet properties that triggers the chain of events leading to the discrete aurora. In view of the particle fluxes encountered in the tail lobes being much lower than in the surrounding regions it may be necessary to mount a dedicated mission into this region. The tail lobes are almost voids, but they have a key role to play.

There is much to suggest that at the Earth's bow shock the same kind of process operates as in the aurora, and the same (lower-hybrid) waves are the agents of transfer of energy from ions (this time of the solar wind) to (solar-wind) electrons. They would certainly appear to operate in the right energy range. The stages of the transition remain highly elusive because of the thinness of the layer and its rapid readjustments of position. However, the often oscillatory motion does sometimes offer a number of crossings and re-crossings which might in principle be re-assembled using the disentangling technique found to order the relatively slow crossings of the magnetopause. Very high rates of data acquisition are required again, especially for the measurements of waves. One of the major uncertainties we encountered in trying to assess exactly what had happened to the electrons in the course of acceleration was the present experimental low energy cut-off at or around 10 eV. We need to know what goes on at least an order of magnitude below this. The difficulties arising from contamination and spacecraft charging need to be overcome before this key region of the electron distribution can be explored. Similar considerations would apply to any future ion-release experiments whether in the solar wind or the magnetosphere.

Difficulties become still greater with solar-wind and solar-flare electrons, for here we have little or no information on the source distributions before acceleration, and scant knowledge of the medium through which they must travel before detection. For cosmic electrons the uncertainty becomes complete. The greatest of all the acceleration mysteries—pertaining to the highest energies and occupying the most extensive region of space—is thus, most fittingly, rendered the most obscure. As an indication of the difficulty of these problems we might consider that it would be difficult enough even in the laboratory to identify the type and internal workings of an accelerator given only the stream of particles emanating from it. It would be next to impossible if there were, in addition, an unknown intervening region of moderating fields. This is not an exhortation to give up—the issues are of too great a significance for that, and must continue to be accepted as a challenge—but it does amount to a plea to keep an open mind and to resist being painted into a corner as the result of mere assumptions, even 'reasonable' ones. On the positive side is the realisation that the medium of outer space is not limited to the properties of a 'hot gas' but rejoices in the far greater capabilities and complexities of collisionless plasmas.

Appendix 1. QBasic program for studying evolution of an energy distribution subject to entropy-promoting internal exchanges of energy

In the program listing below the symbol >> at the end of a line indicates that the material in the line below should follow on when re-typing the code.

```
' AP1.BAS
' Program for studying evolution of an arbitrarily chosen energy
' distribution subject to entropy-promoting internal exchanges of
' energy.

                   CLS
'=====================================================================
' Array dimensions
         DIM F(21, 5)        'energy distribution
         DIM ENT(5)          'entropy
         DIM NUM(5)          'total number
         DIM ENG(5)          'total energy


'  ====================================================================
' Control parameters
         RANDOMIZE 123456    ' random number origin
         jumpx% = 2          ' maximum energy jump (suggest 1,2 or 3)
         nstage% = 4         ' number of stages (1 to 4)
         nmax% = 10000       ' number of attempted operations per stage
         F(1, 1) = 0         ' initial distribution at zero energy
         F(2, 1) = 10000     ' initial distribution at 1 unit of energy
         F(3, 1) = 0         ' initial distribution at 2 units of energy
         F(4, 1) = 0         ' etc.
         F(5, 1) = 0
         F(6, 1) = 0
```

```
      F(7, 1) = 0
      F(8, 1) = 0
      F(9, 1) = 0
      F(10, 1) = 0
      F(11, 1) = 0
      F(12, 1) = 0
      F(13, 1) = 0    ' screen listing stops here
      F(14, 1) = 0
      F(15, 1) = 0
      F(16, 1) = 0
      F(17, 1) = 0
      F(18, 1) = 0
      F(19, 1) = 0
      F(20, 1) = 0
      F(21, 1) = 0
' ================================================================

                IF nstage% > 4 THEN
                nstage% = 4
                ELSE
                END IF
PRINT , "Evolution of energy distribution under increasing entropy"

                jmax% = nstage% + 1
                j% = 1                    ' stage counter
10              j% = j% + 1
                IF j% > jmax% THEN
                GOTO 20                   ' .....collect and print results
                ELSE

' distribution becomes input to new stage
                FOR i% = 1 TO 21
                F(i%, j%) = F(i%, j% - 1)
                NEXT i%

' Commentary on progress
                PRINT "computing stage"; j% - 1
                FOR n% = 1 TO nmax%

' count particles
                s = 0
                FOR M% = 1 TO 21
                s = s + F(M%, j%)
                NEXT M%

'  select one at random
                R = s * RND
```

```
'   find which on which energy step the randomly selected particle
'   resides
                 s = 0
                 FOR M% = 1 TO 21
                 s = s + F(M%, j%)
                 IF R < s THEN
                 k% = M% ' energy step found
                 GOTO 30
                 ELSE
                 END IF
                 NEXT M% ' top energy step - cannot proceed - try next

' ensure that there are at least two particles on the step
30               IF F(k%, j%) < 2 THEN
                 GOTO 40 ' too few particles, abort + make new attempt
                 ELSE

' randomly select the size of jump
                 jump% = INT(RND * jumpx%) + 1
                 END IF

' ensure that step is not too close to an end of the distribution
                 IF k% + jump% < 21 AND k% > jump% THEN
' go ahead
                 GOTO 50

' abort and try next
                 ELSE
                 GOTO 40
                 END IF

' establish provisional modified numbers
50               FT1 = F(k%, j%) - 2
                 FT2 = F(k% - jump%, j%) + 1
                 FT3 = F(k% + jump%, j%) + 1

' evaluate magnitude of components of modified  entropy
                 A = -FT1 * LOG(FT1 + 1)
                 B = -FT2 * LOG(FT2 + 1)
                 C = -FT3 * LOG(FT3 + 1)

' evaluate magnitude of components of un-modified entropy
                 D = -F(k%, j%) * LOG(F(k%, j%) + 1)
                 E = -F(k% - jump%, j%) * LOG(F(k% - jump%, j%) + 1)
                 F = -F(k% + jump%, j%) * LOG(F(k% + jump%, j%) + 1)

' find whether entropy would be increased by the proposed changes
                 IF A + B + C > D + E + F THEN
```

```
' it would; therefore make the changes
            F(k%, j%) = FT1
            F(k% - jump%, j%) = FT2
            F(k% + jump%, j%) = FT3
            ELSE
            GOTO 40
            END IF

' select another in either case
40          NEXT n%

' is there another stage to process?
            IF j% = jmax% THEN
            GOTO 20
            ELSE
            END IF

' there is at least one more stage
' use the modified distribution as initial one for the next stage
' begin next stage
            GOTO 10
            END IF

' begin next stage
            GOTO 10

' collect results
' evaluate entropy, total energy and total number for each stage,
' including initial one

20          FOR j% = 1 TO 5
            ENT(j%) = 0
            ENG(j%) = 0
            NUM(j%) = 0
            FOR i% = 1 TO 21
            ENT(j%) = ENT(j%) - F(i%, j%) * LOG(F(i%, j%) + 1)
            ENG(j%) = ENG(j%) + (i% - 1) * F(i%, j%)
            NUM(j%) = NUM(j%) + F(i%, j%)
            NEXT i%
            ENT(j%) = INT(ENT(j%))
            NEXT j%

' evaluate overall "exposure", ie mean number of encounters per
' particle
            V = nmax%
            W = nstage%
            X = 2 * (V / NUM(1)) * W
```

```
' display results on screen
                CLS
PRINT , "Evolution of energy distribution under increasing entropy"
PRINT
PRINT , "encounters/stage"; USING "######"; nmax%;
PRINT , "jump"; USING "###"; jumpx%;
PRINT , "exposure"; USING "#####.##"; X
PRINT , "      Energy              Number after stage"
PRINT , "             Original   1      2      3      4"
PRINT , " "
                FOR i% = 1 TO 13  'display energies 0-12 only
PRINT , USING "########"; i% - 1; F(i%, 1); F(i%, 2); F(i%, 3); F(i%, >>
  4); F(i%, 5)
                NEXT i%
PRINT , "entropy"; USING "########"; TAB(23); ENT(1); ENT(2); ENT(3); >>
ENT(4); ENT(5)
PRINT , "number "; USING "########"; TAB(23); NUM(1); NUM(2); NUM(3); >>
NUM(4); NUM(5)
PRINT , "energy "; USING "########"; TAB(23); ENG(1); ENG(2); ENG(3); >>
ENG(4); ENG(5)
                STOP
                END
```

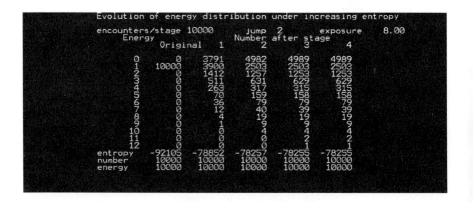

Figure A1. Program output.

Appendix 2. QBasic program for studying evolution of an energy distribution subject to random internal exchanges of energy

In the program listing below the symbol >> at the end of a line indicates that the material in the line below should follow on when re-typing the code.

```
'AP2.BAS
' Program for studying evolution of an energy distribution subject to
' random internal exchanges of energy.

                CLS
'================================================================
' Array dimensions
DIM F(21, 5)            'energydistribution
DIM ENG(5)              'total energy
DIM ENT(5)              'entropy
DIM NUM(5)              'total number
DIM K%(2)               'selected-particle index


'  ==============================================================
' Control parameters
RANDOMIZE 123456        'random number origin
jumpx% = 4              'maximum energy jump (suggest 1 to 3)
stage% = 4              'number of stages (1 to 4)
nmax% = 10000           'number of particle-particle encounters per
                        'stage

' choose initial distribution
        F(1, 1) = 3000    'initial distribution at zero energy
        F(2, 1) = 7000    'initial distribution at 1 unit of energy
        F(3, 1) = 1000    'initial distribution at 2 units of energy
        F(4, 1) = 100     'etc
        F(5, 1) = 10
```

```
                F(6, 1) = 1
                F(7, 1) = 0
                F(8, 1) = 0
                F(9, 1) = 0
                F(10, 1) = 0
                F(11, 1) = 0
                F(12, 1) = 0
                F(13, 1) = 0        'screen listing stops at 12 units of energy
                F(14, 1) = 0
                F(15, 1) = 0
                F(16, 1) = 0
                F(17, 1) = 0
                F(18, 1) = 0
                F(19, 1) = 0
                F(20, 1) = 0
                F(21, 1) = 0
'==================================================================
' Computation
                IF nstage% > 4 THEN
                nstage% = 4
                ELSE
                END IF
PRINT "Evolution of energy distribution by random internal energy"
PRINT "exchange"
                jmax% = stage% + 1        ' stage counter

' count particles
                total = 0
                FOR M% = 1 TO 21
                total = total + F(M%, 1)
                NEXT M%

' start stage counter
                j% = 1

' increment stage counter
10              j% = j% + 1
                IF j% > jmax% THEN
                GOTO 20                   ' collect and print results
                ELSE

' comment on progress
PRINT "computing stage"; j% - 1

' distribution for previous stage becomes input for this stage
                FOR i% = 1 TO 21
                F(i%, j%) = F(i%, j% - 1)
                NEXT i%
```

```
' begin processing
                FOR N% = 1 TO nmax%

' select two particles at random
                FOR index% = 1 TO 2

'               select a particle at random from the total
                R = total * RND

'               search for step on which particle lies
'               zero sum
                s = 0
                FOR M% = 1 TO 21
                s = s + F(M%, j%)
                IF R < s THEN
                K%(index%) = M% ' particle's step found
                GOTO 30            ' see if there is another to find
                ELSE
                GOTO 40            'see if scan can continue
                END IF
40              NEXT M%            ' continue scan
30              NEXT index%        ' go to second particle or, if found,
                                   ' proceed

' put the two selected particles in velocity order
                IF K%(1) < K%(2) THEN
                K1% = K%(1)
                K2% = K%(2)
                ELSE
                K1% = K%(2)
                K2% = K%(1)
                END IF

' K1% is now less than or equal to K2%

' ensure that particles  will not jump below lowest step
                IF K1% > 1 THEN
                GOTO 50            ' on lowest step
                ELSE               ' one or both on lowest step
                IF K1% = 1 THEN
                IF K2% < 3 THEN    ' both on two lowest steps -
                                   ' leave unchanged
                GOTO 60            ' try next
                ELSE               ' k1=1 and k2 = 3 or higher

'jump can be only in one sense,
' so execute on only 50% of occasions to avoid bias
```

```
                p% = INT(RND + .5)          ' p%  0 or 1
                IF p% = 0 THEN              'make jumps

'lower jumps from step 1 to step 2
                F(1, j%) = F(1, j%) - 1
                F(2, j%) = F(2, j%) + 1

' upper jumps down a step
                F(K2%, j%) = F(K2%, j%) - 1
                F(K2% - 1, j%) = F(K2% - 1, j%) + 1
                ELSE              ' do not make jumps
                END IF
                GOTO 60          ' try next
                END IF
                ELSE
                GOTO 60
                END IF          ' for k1 =1
50              IF K1% = K2% THEN
                IF F(K1%, j%) < 2 THEN
                GOTO 60          ' same particle selected twice (!) -
                                 ' try next
                ELSE

' one jumps up, one jumps down, source step loses 2
                F(K1%, j%) = F(K1%, j%) - 2
                F(K1% - 1, j%) = F(K1% - 1, j%) + 1
                F(K1% + 1, j%) = F(K1% + 1, j%) + 1
                GOTO 60    ' try next
                END IF
                ELSE

' general mid-range case
' randomly select size of jump
                jump% = INT(RND * (jumpx% + 1))

' randomly select direction of jump
                p% = INT(RND + .5)  ' p%  0 or 1
                p% = (-1) ^ p%      ' change to p% +1 for lower to
                                    ' jump up, and upper down
                                    ' or  p% = -1 for lower to
                                    ' jump down, and upper up
                IF K2% - K1% = 1 AND p% = 1 THEN ' particles simply
                                     ' interchange
                GOTO 60                          'try next
                ELSE
                IF K2% - K1% = 2 AND p% = 1 THEN
```

```
' both jump onto intervening step
                F(K1%, j%) = F(K1%, j%) - 1
                F(K1% + 1, j%) = F(K1% + 1, j%) + 2
                F(K2%, j%) = F(K2%, j%) - 1
                GOTO 60                    'try next
                ELSE

' if, necessary, curtail size of jump
                IF jump% > K1% - 1 THEN
                jump% = K1% - 1
                ELSE
                END IF

' one jumps up, one jumps down, depending on sign of p%
                y% = p% * jump%
                F(K1%, j%) = F(K1%, j%) - 1
                F(K2%, j%) = F(K2%, j%) - 1
                F(K1% + y%, j%) = F(K1% + y%, j%) + 1
                F(K2% - y%, j%) = F(K2% - y%, j%) + 1
                GOTO 60
                END IF
                END IF
                END IF
                END IF
60              NEXT N%
                GOTO 10
                END IF
                GOTO 10
20              FOR j% = 1 TO 5
                FOR i% = 1 TO 21
                ENT(j%) = ENT(j%) - F(i%, j%) * LOG(F(i%, j%) + 1)

' (the addition of the"1" is a subterfuge
'                         to exclude zeros from entropy evaluation )
                ENG(j%) = ENG(j%) + (i% - 1) * F(i%, j%)
                NUM(j%) = NUM(j%) + F(i%, j%)
                NEXT i%
                NEXT j%

'evaluate overall exposure, ie mean number of encounters per particle
                V = nmax%
                W = stage%
                X = 2 * (V / NUM(1)) * W

' Display results on screen
                CLS
PRINT ,  "Evolution of energy distribution under random internal";
PRINT "energy exchange"
```

```
PRINT "   encounters/stage"; USING "######"; nmax%;
PRINT , "jump"; USING "###"; jumpx%;
PRINT , "exposure"; USING "#####.###"; X
PRINT , "     Energy                Number after stage "
PRINT , "              Original    1       2       3       4 "
PRINT

                FOR i% = 1 TO 13      ' for energies 1 to 12 only
PRINT , USING "########"; i% - 1; F(i%, 1); F(i%, 2); F(i%, 3); F(i%, >>
 4); F(i%, 5)
                NEXT i%
PRINT , " entropy"; USING "########"; TAB(23); ENT(1); ENT(2); ENT(3) >>
; ENT(4); ENT(5)
PRINT , " energy "; USING "########"; TAB(23); ENG(1); ENG(2); ENG(3) >>
; ENG(4); ENG(5)
PRINT , " number "; USING "########"; TAB(23); NUM(1); NUM(2); NUM(3) >>
; NUM(4); NUM(5)
PRINT
                STOP
                END
```

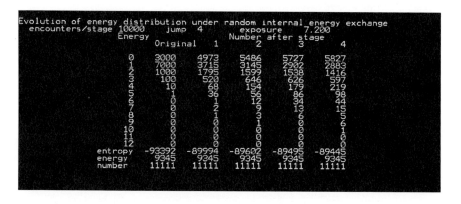

Figure A2. Program output.

Appendix 3. QBasic program for studying evolution of an energy distribution whose particles are subject to random energy exchanges with waves

In the program listing below the symbol >> at the end of a line indicates that the material in the line below should follow on when re-typing the code.

```
' AP3.BAS
' Qbasic program for studying evolution of a particle velocity
' distribution subject to random enrgy exchanges with waves.
                CLS
'====================================================================
' Array dimensions
        DIM F(22, 5)     ' particle velocity distribution
        DIM ENG(5)       ' total energy
        DIM ENT(5)       ' entropy
        DIM NUM(5)       ' total number of particles
        DIM W(42)        ' relative wave- velocity distribution

'====================================================================
' Control parameters
        RANDOMIZE 123456 'base for random number selection
        jumpx% = 2    ' tolerance on resonance
                      ' (measure of wave amplitude)
        stages% = 4   ' number of stages (1 to 4)
        nmax% = 1000  ' number of encounters per stage
                      ' (measure of wave intensity)
        choice = 2    ' choice = 1 for tailored initial particle
                      ' distribution
                      ' choice = 2 for computed power law
        GAMMA = -3    ' exponent of computed power law
        C = 1000      ' magnitude of computed power-law at 1 unit of
```

269

```
                              ' velocity

' Tailored initial particle velocity distribution
                    F(1, 1) = 5000
                    F(2, 1) = 1000
                    F(3, 1) = 500
                    F(4, 1) = 100
                    F(5, 1) = 50
                    F(6, 1) = 10
                    F(7, 1) = 0
                    F(8, 1) = 0
                    F(9, 1) = 0
                    F(10, 1) = 0
                    F(11, 1) = 0
                    F(12, 1) = 0
                    F(13, 1) = 0
                    F(14, 1) = 0
                    F(15, 1) = 0
                    F(16, 1) = 0
                    F(17, 1) = 0
                    F(18, 1) = 0
                    F(19, 1) = 0
                    F(20, 1) = 0
                    F(21, 1) = 0

' Computed power law
                    IF choice > 1 THEN
                    FOR i% = 1 TO 21
                    F(i%, 1) = INT(C * i% ^ GAMMA)
                    NEXT i%
                    ELSE
                    END IF

' enter wave distribution at half-incremental values
                    W(1) = 0     ' relative number of waves of
                                 ' phase velocity 1.5
                    W(2) = 0     ' relative number of waves of
                                 ' phase velocity 2
                    W(3) = 1000 ' at 2.5
                    W(4) = 0     ' at 3
                    W(5) = 0     ' etc.
                    W(6) = 0
                    W(7) = 0
                    W(8) = 0
                    W(9) = 0
                    W(10) = 0
                    W(11) = 0
                    W(12) = 0
```

```
                    W(13) = 0
                    W(14) = 0
                    W(15) = 0
                    W(16) = 0
                    W(17) = 0
                    W(18) = 0
                    W(19) = 0
                    W(20) = 0
                    W(21) = 0
                    W(22) = 0
                    W(23) = 0
                    W(24) = 0
                    W(25) = 0
                    W(26) = 0
                    W(27) = 0
                    W(28) = 0
                    W(29) = 0
                    W(30) = 0
                    W(31) = 0
                    W(32) = 0
                    W(33) = 0
                    W(34) = 0
                    W(35) = 0
                    W(36) = 0
                    W(37) = 0
                    W(38) = 0
                    W(39) = 0 ' 20.5 highest effective wave velocity
                    W(40) = 0
'=======================================================================
' Computation
' count particles
                    sp = 0
                    FOR M% = 1 TO 21
                    sp = sp + F(M%, 1)
                    NEXT M%

' count waves
                    SW = 0
                    FOR M% = 1 TO 40
                    SW = SW + W(M%)
                    NEXT M%

PRINT "Evolution of particle velocity distribution exposed to waves"
                    jmax% = stages% + 1 ' stage index

' begin stage 1
                    j% = 1
10                  j% = j% + 1 'increment stage index
```

```
' have all stages been processed?
                IF j% > jmax% THEN
                GOTO 20   ' Yes, so assemble results
                ELSE      ' No, so begin first or next stage
                END IF
' make current distribution input for first or next stage
                FOR i% = 1 TO 21
                F(i%, j%) = F(i%, j% - 1)
                NEXT i%


' commentary on progress
                PRINT "computing stage "; j% - 1

                FOR n% = 1 TO nmax%

' select a particle at random
                R = sp * RND

' find its velocity (determined by its position in the distribution)
                S = 0
                FOR M% = 1 TO 21
                S = S + F(M%, j%)
                IF R < S THEN
                k% = M% '    Particle velocity found
                GOTO 30
                ELSE
                END IF
                NEXT M%

' select a wave at random
30              R = SW * RND

' find its velocity
                S = 0
                FOR M% = 1 TO 40
                S = S + W(M%)
                IF R < S THEN
                kw = M% ' wave velocity selected
                GOTO 40
                ELSE
                END IF
                NEXT M%

' compute velocity difference,
' acknowledging that waves are spaced at half integer velocities
40              diff = .5 * kw + 1 - k%
```

```
' find whether wave of maximun possible amplitude
' would be large enough to cope with velocity difference
              IF diff * diff > jumpx% * jumpx% THEN
              GOTO 50 ' resonance impossible, try next
              ELSE     ' resonance possible
              END IF

' ensure that there is a particle at the randomly selected velocity
              IF F(k%, j%) < 1 THEN
              GOTO 50  ' no particles, so try next
              ELSE

' randomly select wave "amplitude" in terms of velocity-changing
' capability
              amp = INT(RND * jumpx%) + 1
              END IF

' find index of prospective new particle velocity
              ind% = k% + 2 * diff ' prospective new velocity index

' ensure that it lies within experiment range
              IF ind% < 22 AND ind% > 0 THEN
              GOTO 60 ' jump would remain in range
              ELSE
              GOTO 50 ' jump would be out of range, try next
              END IF

' ensure that wave is large enough to span velocity difference
60            IF diff * diff > amp * amp THEN

              GOTO 50 ' non-resonant try next
              ELSE
              GOTO 70 ' resonant
              END IF

' check that source and target velocities do not have equal numbers
' already
70            IF F(k%, j%) = F(ind%, j%) THEN
              GOTO 50 ' equal already, try next
              ELSE
              END IF

' move particle from source to target velocity
              F(k%, j%) = F(k%, j%) - 1
              F(ind%, j%) = F(ind%, j%) + 1

50  NEXT n%
              GOTO 10 ' to begin next stage
```

```
' compute entropy, energy (proportional to velocity squared),
' and total numbers
20                FOR j% = 1 TO 5
                  FOR i% = 1 TO 21
                  ENT(j%) = ENT(j%) - F(i%, j%) * LOG(F(i%, j%) + 1)
                  ENG(j%) = ENG(j%) + i% ^ 2 * F(i%, j%)
                  NUM(j%) = NUM(j%) + F(i%, j%)
                  NEXT i%
                  NEXT j%


'compute overall exposure
                  V = nmax%
                  W = stages%
                  X = (V / NUM(1)) * W
PRINT V; W; NUM(1); X


' display results on screen
                  CLS
PRINT , "Evolution of particle velocity distribution exposed to waves >
"
PRINT , "  encounters/stage"; USING "######"; nmax%;
PRINT , "jump"; USING "###"; jumpx%;
PRINT , "exposure"; USING "#####.###"; X
PRINT , "    Velocity              Number after stage        waves"
PRINT "                          original     1       2       3     4  >:
    int half-int "


                  i% = 1
PRINT , USING "########"; i%; F(i%, 1); F(i%, 2); F(i%, 3); F(i%,4);  >>
F(i%, 5); 0; W(1)


                  FOR i% = 2 TO 12
PRINT , USING "########"; i%; F(i%, 1); F(i%, 2); F(i%, 3); F(i%, 4); >>
F(i%, 5); W(2 * i% - 2); W(2 * i% - 1)


                  NEXT i%
PRINT , " entropy"; USING "########"; ENT(1); ENT(2); ENT(3); ENT(4); >>
ENT(5)
PRINT , " energy "; USING "########"; ENG(1); ENG(2); ENG(3); ENG(4); >>
ENG(5)
PRINT , " number "; USING "########"; NUM(1); NUM(2); NUM(3); NUM(4); >>
NUM(5)


                  STOP
                  END
```

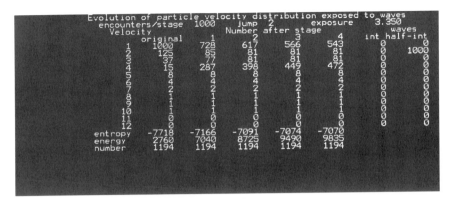

Figure A3. Program output.

Appendix 4. Key formulae

All quantities are in (unprefixed) Standard International (SI) units except where indicated. Open square brackets indicate preferred derivatives, i.e.

$E[\ ]$ = kinetic energy in eV

$j[\ ]$ = differential intensity in m^{-2} s^{-1} sr^{-1} keV^{-1}

$f[\ ]$ = velocity-space density in km^{-6} s^3

$F[\ ]$ = phase-space density in $(zJ\ s)^{-3}$

Velocity

$$v = p/\Gamma m \qquad\qquad m\ s^{-1}$$
$$v_{NR} = p/m \qquad\qquad m\ s^{-1}$$
$$v_{UR} = c \qquad\qquad m\ s^{-1}$$

Pitch angle

$$\alpha = \tan^{-1}(v_\perp/v_\parallel) \qquad\qquad \text{radians or degrees}$$

Relative velocity

Velocity u measured from frame moving at v (u and v measured in same sense)

$$v_{rel} = (u - v)/(1 - uv/c^2)$$

$$v_{rel\,NR} = u - v$$

If either u or v or both = c

$$v_{rel\,UR}^2 = c^2$$

Relativistic (Lorentz) factor

$$\Gamma \equiv 1/\sqrt{1 - v^2/c^2}$$
$$= 1/\sqrt{1 - \beta^2}$$
$$= 1 + \varepsilon$$
$$\Gamma_{NR} = 1 + \varepsilon$$
$$\Gamma_{UR} = \varepsilon$$

Lorentz force

$$\mathcal{F} = Ze(E + vB \sin \alpha) \qquad \text{N}$$
$$\mathcal{F}_e = -1.60 \times 10^{-19}(E + vB \sin \alpha) \qquad \text{N}$$

Acceleration

General

$$a = dv/dt \qquad \text{m s}^{-2}$$
$$= \mathcal{F}/m \qquad \text{m s}^{-2}$$

Radiation by accelerated charged particle

Electronic charge

$$P = Z^2 e^2 a^2/6\pi \varepsilon_0 c^3 \qquad \text{W}$$
$$= 5.7 \times 10^{-54} Z^2 a^2 \qquad \text{W}$$
$$P_e = 5.7 \times 10^{-54} a^2 \qquad \text{W}$$

Momentum

$$p = mc\sqrt{\Gamma^2 - 1} \qquad \text{N s}$$
$$= (E_0/c)\sqrt{\Gamma^2 - 1} \qquad \text{N s}$$
$$= (1/c)\sqrt{E^2 + 2EE_0} \qquad \text{N s}$$
$$p_{NR} = mv \qquad \text{N s}$$
$$p_{UR} = \Gamma mc \qquad \text{N s}$$
$$p_e = 0.273\sqrt{\Gamma^2 - 1} \qquad \text{zN s}$$
$$= 5.34 \times 10^{-7}\sqrt{E^2[\ ] + 2E[\ ]E_0[\ ]} \qquad \text{zN s}$$

$$p_{e\,NR} = 9.11 \times 10^{-10} v \qquad\qquad \text{zN s}$$

$$= 5.40 \times 10^{-4} \sqrt{E[\ ]} \qquad \text{zN s}$$

$$p_{e\,UR} = 0.273\varepsilon \qquad\qquad \text{zN s}$$

Energy

Rest energy

$$E_0 = mc^2 \qquad\qquad \text{eV}$$

$$E_{0e} = 5.11 \times 10^5 \qquad\qquad \text{eV}$$

Kinetic energy

$$E = (\Gamma - 1)mc^2 \qquad\qquad \text{J}$$

$$E_{NR} = \tfrac{1}{2}mv^2 \qquad\qquad \text{J}$$

$$E_{UR} = \Gamma mc^2 \qquad\qquad \text{J}$$

$$E_e = 5.11 \times 10^5(\Gamma - 1) \qquad \text{eV}$$

$$E_{e\,NR} = 2.84 \times 10^{-12} v^2 \qquad \text{eV}$$

$$E_{e\,UR} = 5.11 \times 10^5 \Gamma \qquad\qquad \text{eV}$$

Volume in real space

$$V = \delta x \delta y \delta z \qquad\qquad \text{m}^3$$

Volume in velocity space

$$\mathcal{V} = \delta v_x \delta v_y \delta v_z \qquad\qquad \text{m}^3\ \text{s}^{-3}$$

$$= \mathcal{V}/m^3(1+\varepsilon)^5 \qquad \text{m}^3\ \text{s}^{-3}$$

Volume in real space and velocity space combined

$$\mathcal{T} = V\mathcal{V} \qquad\qquad \text{m}^6\ \text{s}^{-3}$$

Volume in momentum space

$$\mathcal{V} = \delta p_x \delta p_y \delta p_z \qquad\qquad (\text{N s})^3$$

$$= m^3(1+\varepsilon)^5 \mathcal{V} \qquad\qquad (\text{N s})^3$$

Volume in phase space

$$\tau = V\mathcal{V} \qquad \text{(J s)}^3$$

Number density

$$n = N/V \qquad \text{m}^{-3}$$

Velocity-space density

$$f = N/\tau \qquad \text{m}^{-6}\,\text{s}^3$$
$$= Fm^3(1+\varepsilon)^5 \qquad \text{m}^{-6}\,\text{s}^3$$
$$= jm^3(1+\varepsilon)^5/p^2 \qquad \text{m}^{-6}\,\text{s}^3$$
$$= jm^2(1+\varepsilon)^5/\{E(2+\varepsilon)\} \qquad \text{m}^{-6}\,\text{s}^3$$
$$f_{\text{NR}} = jm^2/2E \qquad \text{m}^{-6}\,\text{s}^3$$
$$= Fm^3 \qquad \text{m}^{-6}\,\text{s}^3$$
$$f_{\text{e}} = 7.56 \times 10^{-10} F_e[\ ](1+\varepsilon)^5 \qquad \text{km}^{-6}\,\text{s}^3$$
$$= 3.23 \times 10^{-8} j_e[\ ](1+\varepsilon)^5/\{E[\ ](2+\varepsilon)\} \qquad \text{km}^{-6}\,\text{s}^3$$
$$f_{\text{e\,NR}} = 7.56 \times 10^{-10} F_e[\ ] \qquad \text{km}^{-6}\,\text{s}^3$$
$$= 1.62 \times 10^{-8} j_e[\ ]/E[\ ] \qquad \text{km}^{-6}\,\text{s}^3$$

Phase-space density

$$F = N/\tau \qquad \text{(J s)}^{-3}$$
$$F = j/p^2 \qquad \text{(J s)}^{-3}$$
$$F_{\text{NR}} = j/2mE \qquad \text{(J s)}^{-3}$$
$$= f_{\text{NR}}/m^3 \qquad \text{(J s)}^{-3}$$
$$F_{\text{UR}} = jc^2/E^2 \qquad \text{(J s)}^{-3}$$
$$F = f/\{m^3(1+\varepsilon)^5\} \qquad \text{(J s)}^{-3}$$
$$F_{\text{e}} = 1.32 \times 10^9 f_e[\ ]/(1+\varepsilon)^5 \qquad \text{(zJ s)}^{-3}$$
$$= 2.19 \times 10^7 j_e[\ ]/\{E^2[\ ](1+2/\varepsilon)\} \qquad \text{(zJ s)}^{-3}$$
$$F_{\text{e\,NR}} = 1.32 \times 10^9 f_e[\ ] \qquad \text{(zJ s)}^{-3}$$
$$= 21.4 j_e[\ ]/E[\ ] \qquad \text{(zJ s)}^{-3}$$
$$F_{\text{eUR}} = 2.19 \times 10^7 j_e[\ ]/E^2[\ ] \qquad \text{(zJ s)}^{-3}$$

Differential intensity

$$j = p^2 F \qquad\qquad \text{m}^{-2}\ \text{s}^{-1}\ \text{sr}^{-1}\ \text{J}^{-1}$$

$$\quad = N/\mathrm{d}A\mathrm{d}t\mathrm{d}\Omega\mathrm{d}E \qquad\qquad \text{m}^{-2}\ \text{s}^{-1}\ \text{sr}^{-1}\ \text{J}^{-1}$$

$$j_e = 4.57 \times 10^{-8} F_e[\ ]E^2[\ ]\{1 + 2/\varepsilon\} \qquad\qquad \text{m}^{-2}\ \text{s}^{-1}\ \text{sr}^{-1}\ \text{keV}^{-1}$$

$$j_{e\,NR} = 4.67 \times 10^{-2} F_e[\ ]E[\ ] \qquad\qquad \text{m}^{-2}\ \text{s}^{-1}\ \text{sr}^{-1}\ \text{keV}^{-1}$$

$$j_{e\,UR} = 4.57 \times 10^{-8} F_e[\ ]E^2[\ ] \qquad\qquad \text{m}^{-2}\ \text{s}^{-1}\ \text{sr}^{-1}\ \text{keV}^{-1}$$

Liouville's theorem

In a dynamical, reversible system

$$F \text{ and } f \text{ remain constant}$$

Extension to a non-dynamical system

$$\mathrm{d}F/\mathrm{d}t = -F\mathrm{d}a/\mathrm{d}v$$

$$\mathrm{d}f/\mathrm{d}t = -f\mathrm{d}a/\mathrm{d}v$$

Entropy

$$S = -k \sum N \ln f + \text{constant}$$

$$(\text{or } S = -k \sum N \ln F + \text{constant})$$

Energy distributions

Maxwellian

$$f_M = n(m/2\pi E_M)^{3/2} \exp\{-E/E_M\} \qquad\qquad \text{m}^{-6}\ \text{s}^3$$

$$\langle E \rangle = 1.5 E_M \qquad\qquad \text{m}^{-6}\ \text{s}^3$$

$$f_{Me} = 0.86n(1/E_M[\ ])^{3/2} \exp\{-E[\ ]/E_M[\ ]\} \qquad\qquad \text{km}^{-6}\ \text{s}^3$$

Power law

$$F \propto E^{-\gamma}$$

(normalization-prevented by divergence of number or energy content)

Kappa

$$f_\kappa = n(m/2\pi\kappa E_\kappa)^{3/2}\frac{\kappa!}{(\kappa-1.5)!}\left(1+\frac{E}{\kappa E_\kappa}\right)^{-(\kappa+1)} \qquad \text{m}^{-6}\,\text{s}^3$$

approximates to Maxwellian for $E \ll \kappa E_\kappa$

approximates to power law for $E \gg \kappa E_\kappa$

$$\langle E\rangle = 1.5\kappa E_\kappa/(\kappa-1.5) \qquad \kappa > 1.5$$

$$f_\kappa \propto (\kappa E_\kappa + E)^{-(\kappa+1)}$$

$$f_{\kappa e} = 0.86n(\kappa E_\kappa[\;])^{-3/2}\frac{\kappa!}{(\kappa-1.5)!}\left(1+\frac{E[\;]}{\kappa E_\kappa[\;]}\right)^{-(\kappa+1)} \qquad \text{km}^{-6}\,\text{s}^3$$

Dynamic scatter

$$F \propto \frac{\left(\sqrt{\varepsilon}+\sqrt{\varepsilon+2}\right)^{-\sqrt{2/N_0\beta^2}}}{\sqrt{\varepsilon}\left(\sqrt{2\varepsilon}+\beta\right)(\varepsilon+2)^2}$$

$$F_{\text{UR}} \propto \varepsilon^{-\left(3+\sqrt{1/2N_0\beta^2}\right)}$$

$\varepsilon \gg \beta \qquad F_{\text{NR}}, f_{\text{NR}} \propto 1/\varepsilon$

$\varepsilon \ll \beta \qquad F_{\text{NR}}, f_{\text{NR}} \propto 1/\sqrt{\varepsilon}$

Temperature

Maxwellian

$$T = 2\langle E\rangle/3k \qquad\qquad \text{K}$$
$$= 7.74\times10^3\langle E[\;]\rangle \qquad \text{K}$$

Other distributions

$$T_{\text{eff}} = 2\langle E\rangle/3k \qquad\qquad \text{K}$$
$$= 7.74\times10^3\langle E[\;]\rangle \qquad \text{K}$$

Debye length

Maxwellian distribution

$$\lambda_{\text{D}} = \sqrt{\varepsilon_0 kT/ne^2} \qquad\qquad \text{m}$$

$$= 69.0\sqrt{T/n} \qquad \text{m}$$

$$= 7.44 \times 10^3\sqrt{E_M[\;]/n} \qquad \text{m}$$

$$= 6.07 \times 10^3\sqrt{\langle E\rangle[\;]/n} \qquad \text{m}$$

Kappa distribution

$$\lambda_\kappa = \sqrt{\frac{\varepsilon_0 \kappa E_\kappa}{(\kappa - 0.5)ne^2}} \qquad \text{m}$$

$$= \sqrt{\frac{2(\kappa - 1.5)\varepsilon_0\langle E\rangle}{3(\kappa - 0.5)ne^2}} \qquad \text{m}$$

$$= 6.07 \times 10^3\sqrt{\frac{(\kappa - 1.5)\langle E\rangle\;[\;]}{(\kappa - 0.5)n}} \qquad \text{m}$$

For same n and $\langle E\rangle$

$$\lambda_\kappa = \lambda_D\sqrt{\frac{\kappa - 1.5}{\kappa - 0.5}}$$

Gyro-radius

$$r_c = p/ZeB \qquad \text{m}$$

$$= 2.08 \times 10^{10}\sqrt{E^2 + 2E_0 E}/ZB \qquad \text{m}$$

$$r_{c\,NR} = 2.94 \times 10^{10}\sqrt{E_0 E}/ZB \qquad \text{m}$$

$$r_{c\,UR} = 2.08 \times 10^{10}E/ZB \qquad \text{m}$$

$$r_{ce} = 3.34 \times 10^{-9}\sqrt{E^2[\;] + 1.02 \times 10^6 E[\;]}/B \qquad \text{m}$$

$$r_{ce\,NR} = 3.37 \times 10^{-6}\sqrt{E[\;]}/B \qquad \text{m}$$

$$r_{ce\,UR} = 3.34 \times 10^{-9}E[\;]/B \qquad \text{m}$$

Mean free path

For randomly placed 'spheres of influence' of radius r

$$\lambda = 1/(\sqrt{2}\pi nr^2)$$

Radial diffusion

$$n = (N/4) (4\pi v\lambda t/3)^{-3/2} \exp\{-3r^2/4v\lambda t\}$$

Gyro angular frequency

$$\omega_c = ZeB/\Gamma m \qquad\qquad \text{rad s}^{-1}$$

$$\omega_{c\,NR} = ZeB/m \qquad\qquad \text{rad s}^{-1}$$

$$\omega_{c\,UR} = ZeBc^2/E \qquad\qquad \text{rad s}^{-1}$$

$$\omega_{ce} = 1.76 \times 10^{11} B/\Gamma \qquad\qquad \text{rad s}^{-1}$$

$$\omega_{ce\,NR} = 1.76 \times 10^{11} B \qquad\qquad \text{rad s}^{-1}$$

$$\omega_{ce\,UR} = 8.99 \times 10^{16} B/E\ [\ ] \qquad\qquad \text{rad s}^{-1}$$

Plasma angular frequency

$$\omega_{pe} = \sqrt{ne^2/m\varepsilon_0} \qquad\qquad \text{rad s}^{-1}$$

$$\omega_{pe} = 56.4\sqrt{n} \qquad\qquad \text{rad s}^{-1}$$

$$\omega_{pi} = \sqrt{ne^2/M\varepsilon_0} \qquad\qquad \text{rad s}^{-1}$$

$$\omega_{pi} = 56.4\sqrt{mn/M} \qquad\qquad \text{rad s}^{-1}$$

Lower-hybrid waves

Wavenumber

$$K = 2\pi/\lambda \qquad\qquad \text{m}^{-1}$$

$$K^2 = K_\perp^2 + K_\parallel^2 \qquad\qquad \text{m}^{-2}$$

Propagation

$$K_\perp/K_\parallel \approx \sqrt{M/m}$$

Angular frequency

$$\omega^2 = \omega_{lh}^2 \left(1 + 3K^2\lambda_D^2 + (K_\parallel/K)^2 M/m\right) \qquad\qquad (\text{rad/s})^2$$

$$\approx \omega_{lh}^2 \left(1 + (K_\parallel/K_\perp)^2 M/m\right) \qquad\qquad (\text{rad/s})^2$$

Resonance frequency

$$\omega_{\mathrm{lh}}^2 = \omega_{\mathrm{pi}}^2/(1 + \omega_{\mathrm{pe}}^2/\omega_{\mathrm{ce}}^2) \qquad\qquad (\mathrm{rad/s})^2$$

Perpendicular phase velocity

$$v_{\phi\perp} = \omega/K_\perp \qquad\qquad \mathrm{m\ s}^{-1}$$

Parallel phase velocity

$$v_{\phi\parallel} = \omega/K_\parallel \qquad\qquad \mathrm{m\ s}^{-1}$$

$$\approx v_{\phi\perp}\sqrt{\frac{M}{m}}\left[\frac{\omega^2}{\omega_{\mathrm{lh}}^2} - 1\right]^{-1/2} \qquad\qquad \mathrm{m\ s}^{-1}$$

Perpendicular group velocity

$$v_{\mathrm{g}\perp} = \mathrm{d}\omega/\mathrm{d}K_\perp \qquad\qquad \mathrm{m\ s}^{-1}$$

$$\approx -v_{\phi\perp}\left[1 - \frac{\omega_{\mathrm{lh}}^2}{\omega^2}\right] \qquad\qquad \mathrm{m\ s}^{-1}$$

Parallel group velocity

$$v_{\mathrm{g}\parallel} = \mathrm{d}\omega/\mathrm{d}K_\parallel \qquad\qquad \mathrm{m\ s}^{-1}$$

$$\approx v_{\mathrm{g}\perp}\sqrt{\frac{M}{m}}\left[\frac{\omega^2}{\omega_{\mathrm{lh}}^2} - 1\right]^{-1/2} \qquad\qquad \mathrm{m\ s}^{-1}$$

$$\approx -v_{\phi\parallel}\left[1 - \frac{\omega_{\mathrm{lh}}^2}{\omega^2}\right] \qquad\qquad \mathrm{m\ s}^{-1}$$

References

[1] Taylor B N 1992 *Phys. Rev.* D **96** (11) part II 1 June 1992, III.1

[2] Fermi E 1949 *Phys. Rev.* **75** 1169, 1049

[3] Faraday M 1832 *Phil. Mag.* **11** 462

[4] Thomson J J 1897 *Phil. Mag.* **41** 293

[5] Van der Graaf R J 1931 *Phys. Rev.* **38** 1919

[6] Alfvén H 1958 *Tellus* **10** 104

[7] Block L P 1978 *Astrophys. Space Sci.* **55** 59

[8] Alfvén H 1981 *Cosmic Plasma* (Dordrecht: Reidel)

[9] Landau L D 1946 J. Phys. (Moscow) **10** 25

[10] Kollath R (ed) 1967 *Particle Accelerators* (London: Pitman) (Engl. Transl.) Greene D 1986 *Linear Accelerators for Radiation Therapy, Medical Physics Handbooks Series* (Bristol: Hilger)

[11] Kerst 1940 *Phys. Rev.* **58** 841; 1942 *Rev. Sci. Instrum.* **13** 387; 1946 *Nature* **157** 90

[12] Swann W F G 1933 *Phys. Rev.* **43** 217

[13] Sweet P A 1958 *Electromagnetic Phenomena in Cosmical Plasmas* ed B Lehnert (Cambridge: Cambridge University Press) p 123

[14] Dungey J W 1958 *Cosmic Electrodynamics* (Cambridge: Cambridge University Press)

[15] Lawrence E O and Livingston M S 1934 *Phys. Rev.* **45** 608

[16] Joos G 1951 *Theoretical Physics* (Glasgow: Blackie)

[17] Davis Jr L 1956 *Phys. Rev.* **101** 351

[18] Bryant D A, Powell G I and Perry C H 1992 *Nature* **356** 582

[19] Vasyliunas V 1968 *J. Geophys. Res.* **77** 2839 who refers to a communication from S Olbert

[20] Swann W F G 1933 *Phys. Rev.* **44** 224

[21] Dendy R O (ed) 1993 *Plasma Physics: An Introductory Course* (Cambridge: Cambridge University Press)

[22] Bryant D A 1996 *J. Plasma Phys.* **56** 87

[23] Bryant D A, Hall D S and Bingham R 1991 *Auroral Physics* ed C-I Meng, M J Rycroft and L A Frank (Cambridge: Cambridge University Press)

[24] Evans D S 1965 *Rev. Sci. Instrum.* **36** 375

[25] Bryant D A and Johnstone A D 1965 *Rev. Sci. Instrum.* **36** 1662

[26] Fraser G W 1989 *X-ray Detectors in Astronomy* (Cambridge: Cambridge University Press)

[27] Kitchin C R 1991 *Astrophysical Techniques* 2nd edn (Bristol: Hilger)

[28] Johnstone A D, Kellock S J, Coates A J, Smith M F, Booker T and Winningham J D 1985 *IEEE Trans. Nucl. Sci.* **NS-32** 139

[29] Johnstone A D 1972 *Rev. Sci. Instrum.* **43** 1030

[30] Staub H 1953 *Experimental Nuclear Physics* vol 1 ed E Segrè (New York: Wiley) p 1

[31] Chaloner C P, Bryant D A and Hall D S 1984 *Adv. Space Sci.* **4** 519

[32] Pease R S 1993 *Plasma Physics: An Introductory Course* ed R O Dendy (Cambridge: Cambridge University Press) p 475

[33] Rairden R L, Frank L A and Craven J D 1986 *J. Geophys. Res.* **91** 13613

[34] Feynman J 1985 *Handbook of Geophysics and the Space Environment* ed A S Jorsa (USAF Geophysics Laboratory Publication)

[35] Faraday M 1852 'On the physical lines of magnetic force' and 'On the lines of magnetic force'. Evening discourses given at the Royal Institution, London, on 11 and 23 June 1852, respectively

    See also, Thomas J M 1991 *Michael Faraday* (Bristol: Hilger)

[36] Parker E N 1963 *Interplanetary Dynamical Processes* (New York: Interscience)

    Hundhausen A J 1972 *Coronal Expansion and Solar Wind* (New York: Springer)

[37] Page D E 1985 *EOS Trans. Am. Geophys. Union* 297 (25 July)

[38] Johnstone A D 1990 *Solar and Planetary Physics* ed B Buti (Singapore: World Scientific) p 209

[39] Ward A K, Bryant D A, Edwards T, Parker D J, O'Hea A, Patrick T J, Sheather P H, Barnsdale K P and Cruise A M 1985 *Geosci. Remote Sensing* **GE-23** 202

[40] Greenstadt E W and Fredericks R W 1979 *Solar System Plasma Physics III* ed C K Kennell, L J Lanzerotti and E N Parker (Amsterdam: North-Holland)

[41] Gold T 1959 *J. Geophys. Res.* **64** 1219; see also 1987 *The Solar Wind and the Earth* ed S-I Akasofu and Y Kamide (Boston, MA/Dordrecht: Terra Scientific/Reidel)

[42] Gombosi T I (ed) 1993 *Plasma Environments of Non-magnetic Planets* (Oxford: Pergamon)

[43] Grewing M, Praderie F and Reinhard R 1988 *Exploration of Halley's Comet* (Berlin: Springer)

[44] Scudder J D, Lind D L and Ogilvie K W 1973 *J. Geophys. Res.* **78** 6535

[45] Foster J C and Burrows J R 1976 *J. Geophys. Res.* **81** 6016

[46] Paschmann G, Baumjohann W, Sckopke N, Phan T-D and Lühr H 1993 *J. Geophys. Res.* **98** 13409

[47] From the AMPTE-UKS electron spectrometer: data courtesy D S Hall

[48] Rossi B 1952 *Cosmic Rays* (New York: McGraw-Hill)

[49] Setti G, Spada G and Wolfendale A W (ed) 1980 *Origin of Cosmic Rays* (Dordrecht: Reidel)

[50] Wilkinson D H 1981 *Prog. Part. Nucl. Phys.* **6** 325

[51] Roederer J G 1970 *Dynamics of Geomagnetically Trapped Radiation* (New York: Springer)

[52] Alfvén H and Fälthamma C-G 1963 *Cosmical Electrodynamics* (Oxford: Clarendon)

    Hastie R J 1993 *Plasma Physics: An Introductory Course* ed R O Dendy (Cambridge: Cambridge University Press) p 5

[53] Ferraro V C 1952 *J. Geophys. Res.* **57** 15

[54] Axford W I and Hines C O 1961 *Can. J. Phys.* **39** 1433

[55] Kivelson M G and Russell C T (ed) 1995 *Introduction to Space Physics* (Cambridge: Cambridge University Press)

[56] Hirshberg J 1969 *J. Geophys. Res.* **74** 5841

[57] Bryant D A 1993 *Plasma Physics: An Introductory Course* ed R O Dendy (Cambridge: Cambridge University Press) p 209

[58] Williams D J, Roelof E C and Mitchell D G 1992 *Rev. Geophys.* **30** 183

[59] Burch J L 1987 *The Solar Wind and the Earth* ed S-I Akasofu and Y Kamide (Tokyo: Terra Scientific) p 103

[60] ESA *International Conference on Substorms, ESA Publication SP-335 (May 1992)* See also, Kennel C F 1995 *Convection and Substorms* (Oxford: Oxford University Press)

[61] Akasofu S-I and Kamide Y 1987 *The Solar Wind and the Earth* ed S-I Akasofu and Y Kamide (Tokyo: Terra Scientific) p 143

[62] Akasofu S-I and Kan J R (ed) 1981 *Physics of Auroral Arc Formation, Geophysical Monograph 25* (Washington, DC: American Geophysical Union)

[63] Winkler J R and Nemzek R J 1993 *Auroral Plasma Dynamics, Geophysical Monograph 80* (Washington, DC: American Geophysical Union) p 1

[64] Akasofu S-I, Meng C-I and Kimball D S 1966 *J. Atmos. Terr. Phys.* **28** 505

[65] Cresswell G R and Davis T N 1966 *J. Geophys Res.* **71** 3155

[66] Bryant D A, Bingham R, Edwards T, Hall D S and Ward A K 1984 *J. Brit. Interplanetary Soc.* **37** 309

[67] Davidson G T 1990 *Space Sci. Rev.* **53** 45

[68] Sandahl I and Eliasson L 1985 *ESA Scientific Publication SP-229* 103

[69] Mitchell D G Williams D J, Huang C Y, Frank L A and Russell C T 1982 *J. Geophys. Res Lett.* **17** 583

[70] Christon S, Williams D, Mitchell F L A and Huang C Y 1989 *J. Geophys. Res.* **94** 13409

[71] Hones Jr E W 1971 *J. Geophys. Res.* **76** 63

[72] Lepine D R, Bryant D A and Hall D S 1980 *Nature* **286** 469

[73] Reasoner D L and Chappell C R 1973 *J. Geophys. Res.* **78** 2176

[74] Reiff P H, Collin H L, Craven J D, Burch J L, Winningham J D, Shelley E G, Frank L A and Freedman M A 1988 *J. Geophys. Res.* **93** 7441

[75] Bryant D A 1981 *Physics of Auroral Arc Formation* ed S-I Akasofu and J Kan *AGU Geophysical Monograph* **25** 103

[76] Egeland A, Maynard N C, Bryant D A, Hall D S, Johnstone A D and Måseide K 1985 *J. Atmos. Terr. Phys.* **47** 693

[77] Winningham J D and Heikkila W J *J. Geophys. Res.* **79** 949

[78] Van Allen J A 1969 *Rev. Geophys.* **7** 233

[79] Singley G W and Vetti J I 1972 *NSSDC 72-06*

[80] Meyer P 1969 *Ann. Rev. Astron. Astrophys.* **7** 1

[81] Nishimura J, Kobayashi T, Komori Y and Yoshida K 1997 *Adv. Space Res.* **19** 767

[82] Lin R P, Mewaldt R A and Van Hollebeke M A 1982 *Astrophys. J.* **235** 949

[83] Bryant D A 1985 *Plasma Phys. Control. Fusion* **27** 1369

[84] Akasofu S-I 1979 *Aurora Borealis; The Amazing Northern Lights Alaska Geographic* **6**(2) (Alaska Geographic Society)

[85] Davis T N, Hallinen T J and Stenbaek-Nielsen H C 1971 *The Radiating Atmosphere* ed B M McCormack (Dordrecht: Reidel) p 160

Belon A E, Maggs J E, Davis T N, Mather K B, Glass N W and Hughes G F 1969 *J. Geophys. Res.* **74** 1

[86] Bone N 1994 *The Aurora* (New York: Wiley-Praxis)

[87] Krukonis A P and Whalen B A 1980 *J. Geophys. Res.* **85** 119

[88] Aurora Color Television Project, Geophysical Institute, University of Alaska, Fairbanks AK 99775-7320 (http://dac3.pfrr.alaska.edu:80/ pfrr/AURORA/)

[89] Bryant D A 1994 *Contemp. Phys.* **35** 165

[90] Swift J 1726 *Gullivers Travels*

[91] Swift D W, and Gorney D J 1989 *J. Geophys. Res.* **94** 2696

[92] Whalen B A and Daly P W 1979 *J. Geophys. Res.* **84** 4175

[93] Arnoldy R L, Lewis P B and Isaacson P O 1974 *J. Geophys. Res* **79** 4208

[94] Evans D S 1975 *Physics of the Hot Plasma in the Magnetosphere* ed B Hultqvist and L Stenflo (New York: Plenum) p 319

[95] Lin C S and Hoffman R A 1979 *J. Geophys. Res.* **84** 6547

[96] McFadden J P, Carlson C W, Boehm M H and Hallinen T J 1987 *J. Geophys. Res.* **92** 11133–48

[97] McFadden J P, Carlson C W and Boehm M H 1986 *J. Geophys. Res.* **91** 1723

[98] McFadden J P, Carlson C W and Boehm M H 1990 *J. Geophys. Res.* **95** 6533

[99] Bryant D A 1987 *ESA Scientific Publication SP-270* p 273

[100] Bryant D A, Hall D S, Lepine D R and Mason R W N 1977 *Nature* **266** 148
See also, Bryant D A 1983 *High Latitude Space Plasma Physics* ed B Hultqvist and T Hagfors (New York: Plenum) p 295

[101] Frank L A and Ackerson K L 1971 *J. Geophys. Res.* **76** 3612
See also, numerous rocket and satellite traversals of auroral zone in Akasofu S-I and Kan J R (ed) 1981 *Physics of Auroral Arc Formation 2, Geophysical Monograph 25* (Washington, DC: American Geophysical Union)
Hultqvist B and Hagfors T (ed) 1982 *High latitude Space Plasma Physics* (New York: Plenum)
Meng C-I, Rycroft M J and Frank L A (ed) 1991 *Auroral Physics* (Cambridge: Cambridge University Press)
Lysak R L (ed) 1993 *Auroral Plasma Dynamics, Geophysical Monograph 80* (Washington, DC: American Geophysical Union)

[102] Courtier G M, Smith M J and Bryant D A 1974 *Magnetospheric Physics* ed B M McCormac (Dordrecht: Reidel) p 207

[103] Lepine D R, Bryant D A and Hall D S 1979 *Proc. 7th Ann. Meeting on Upper Atmosphere Studies by Optical Methods (Tromso Auroral Observatory)* ed O Harang and K Henriksen p 8 (ISSN 0373-4854)

[104] Kaufmann R L and Ludlow G R 1981 *J. Geophys. Res.* **86** 7577

[105] Klumpar D M and Heikkila W J 1982 *Geophys. Res. Lett.* **9** 873

[106] Carlson C W *et al* 1998 *Geophys. Res. Lett.* **25** 2017

[107] Ergun R E *et al* 1998 *Geophys. Res. Lett.* **25** 2025

[108] Sharp R D, Shelley E G and Johnson R G 1980 *J. Geophys. Res.* **85** 92

[109] Menietti J D and Burch J L 1985 *J. Geophys. Res.* **90** 5345

[110] Eliasson L, André M, Lundin R, Pottelette R and Marklund G 1996 *J. Geophys. Res.* **101** 13225

[111] Alfvén H 1939 *Kungl. Sv. Vetenskapsaked Handl.* **18**(3); 1940 **18**(9)

[112] Cowling T G 1942 *Terr. Mag. Atmos. Elec.* **47** 209

[113] Borovsky J E 1992 *Phys. Rev. Lett.* **69** 1054

[114] Bryant D A, Bingham R and de Angelis U 1991 *Phys. Rev. Lett.* **68** 37

[115] McIlwain C E 1960 *J. Geophys. Res.* **65** 2727

[116] Albert R D 1967 *Phys. Rev. Lett.* **18** 369

[117] Evans D S 1968 *J. Geophys. Res.* **73** 2315

[118] Mozer F S, Cattell C A, Hudson M R, Lysak R L, Temerin M and Torbert R B 1980 *Space Sci. Rev.* **27** 155

[119] Baker D N, Chin G and Pfaff Jr R F 1992 *Physics Today* **45** 118
See comment by Bryant D A 1992 *Physics Today* **45** 117 and reply by Baker *et al*, p 119 of same issue

[120] Carlson C W, Pfaff R F and Watzin J G 1998 *Geophys. Res. Lett.* **25** 2013

[121] Lundin R and Eliasson L 1991 *Ann. Geophys.* **9** 202
Also, see comment by Bryant D A 1991 *Ann. Geophys.* **9** 224

[122] Bryant D A 1991 *Ann. Geophys.* **9** 13829

[123] Evans D S 1974 *J. Geophys. Res.* **79** 2853

[124] Pulliam D M, Anderson H R, Stamnes K and Rees M H 1981 *J. Geophys. Res.* **86** 2397

[125] Bryant D A 1976 *The Scientific Satellite Programme during the International Magnetospheric Study* ed K Knott and B Battrick (Dortrecht: Reidel) p 413

[126] Oscarsson T E and Rönmark K G 1990 *Preprint 115* Swedish Institute of Physics

[127] Temerin M, Carlson C and McFadden J P 1993 *Auroral Plasma Dynamics, Geophysical Monograph 155* ed R L Lysak (Washington, DC: American Geophysical Union)

[128] Mälkki A 1993 *EOS Trans. Am. Geophys. Union* **74** 56

[129] Temerin M, Cerny K, Lotko W and Mozer F S 1982 *Phys. Rev. Lett.* **48** 1175

[130] Boström R, Dovner P-O, Eiksson A, Holmgren G and Mälkki A 1993 *Double Layers and other Non-linear Potential Structures in Plasmas* ed R W Schrittwieser (Singapore: World Scientific) p 29

[131] Bryant D A, Bingham R and de Angelis U 1993 *Double Layers and other Non-linear Potential Structures in Plasmas* ed R W Schrittwieser (Singapore: World Scientific) p 42

[132] Stenbaek-Nielsen H C, Hallinen T J, Wescott E M and Foeppl H 1984 *J. Geophys. Res.* **89** 10788
See also, Haerendel G 1987 *The Solar Wind and the Earth* ed S-I Akasofu and Y Kamide (Tokyo: Terra Scientific) p 215

[133] Haerendel H, Rieger E, Valenzuela A, Foeppl H M, Stenbaek-Nielson H C and Wescott E M 1976 *ESA Scientific Publication SP-115* p 203

[134] Wilhelm K 1980 *ESA Scientific Publication SP-152* p 407
Wilhelm K, Bernstein W, Kellogg P J and Whalen B A 1985 *J. Geophys. Res.* **90** 491
Wilhelm K 1985 *ESA Scientific Publication SP-229* p 197

[135] Pottelette R, Malingre M, Bahnsen A, Eliasson L, Stasiewicz K Erlandson P E, and Marklund G 1988 *Ann. Geophys.* **6** 573

[136] Wescott E M, Stenbaek-Nielsen H C, Davis T N, Murcray W H, Peek H M and Bottoms P J 1975 *J. Geophys. Res.* **80** 951
Wescott E M, Stenbaek-Nielsen H C, Davis T N and Peek H M 1976 *J. Geophys. Res.* **81** 4495

[137] Ghielmetti A G, Johnson R G, Sharp R D and Shelley E G 1978 *Geophys. Res. Lett.* **5** 59

Kondo T, Whalen B A, Yau A W and Peterson W K 1990 *J. Geophys. Res.* **95** 12091

[138] Hultqvist B 1987 *Geophys. Res. Lett.* **14** 379

[139] Bryant D A 1992 *Ann. Geophys.* **10** 333

[140] Sharp R D 1981 *Physics of Auroral Arc Formation, Geophysical Monograph 25* ed S-I Akasofu and J Kan (Washington, DC: American Geophysical Union) p 112

[141] McFadden J P *et al* 1998 *Geophys. Res. Lett.* **25** 2021

[142] Collin H L, Sharp R D, Shelley E G and Johnson R G 1981 *J. Geophys. Res.* **86** 6820

[143] Mobius E *et al* 1998 *Geophys. Res. Lett.* **25** 2029

[144] Hultqvist B, Lundin R, Stasiewicz K, Block L, Lindqvist P A, Gustafsson G, Koskinen H, Bahnsen A, Potemra T A and Zanetti L J 1988 *J. Geophys. Res.* **93** 9765

[145] Hultqvist B 1988 *J. Geophys. Res.* **93** 9777

[146] Fälthammar C-G, Block L P, Lindqvist P-A, Marklund G, Pedersen A and Mozer F S 1987 *Ann. Geophys.* A **5** 171

Pottelette R, Malingre M, Bahnsen A, Eliasson L, Stasiewicz K Erlandson P E, and Marklund G 1988 *Ann. Geophys.* **6** 573

[147] Dawson J 1961 *Phys. Fluids* **4** 869

Boyd D A, Stauffer F J, and Trivelpiece A W 1976 *Phys. Rev. Lett.* **37** 98

[148] Chen F E 1974 *Introduction to Plasma Physics* (New York: Plenum)

[149] Stix T 1962 *The Theory of Plasma Waves* (New York: McGraw-Hill) p 224

[150] Cairns R A 1993 *Plasma Physics: An Introductory Course* ed R O Dendy (Cambridge: Cambridge University Press) p 391

[151] Swift D W 1968 *Planet. Space Sci.* **10** 329

[152] Chang T, Crew G R, Retterer J M and Jasperse J R 1989 *IEEE Trans. Plasma Science* **PS-17** 186

[153] Bryant D A 1990 *Solar and Planetary Physics* ed B Buti (Singapore: World Scientific) p 58

[154] Swift D W 1970 *J. Geophys. Res.* **75** 6324

[155] Bryant D A, Hall D S and Lepine D R 1978 *Planet. Space Sci.* **26** 81

[156] Rozmus W and Samson J C 1988 *Phys. Fluids* **31** 2904

[157] Bruhwiler D L and Cary J R 1992 *Phys. Rev. Lett.* **68** 255

[158] Bingham R, Su J J, Dawson J M, McClements K G and Spicer D S 1993 *Astrophys. J.* **409** 465

[159] Bryant D A, Cook A C, Wang Z-S, de Angelis U and Perry C H 1991 *J. Geophys. Res.* **96** 13829

[160] Bryant D A 1990 *Phys. Scripta* T **30** 215–28

[161] Bryant D A and Perry C H 1995 *J. Geophys. Res.* **100** 23711

[162] Stringer T E 1993 *Plasma Physics: An Introductory Course* ed R O Dendy (Cambridge: Cambridge University Press) p 369

[163] Scarf F L, Fredericks R W, Russell C T, Kivelson M, Neugebauer M and Chappell C R 1973 *J. Geophys. Res.* **78** 2150

[164] Vago J L, Kintner P M, Chesney S W, Arnoldy R L, Lynch K A, Moore T E and Pollock C J 1992 *J. Geophys. Res.* **97** 16935

[165] Pécseli H L, Iranpour K, Holter Ø, Lybekk B, Holtet J, Trulsen J, Eriksson A and Holback B 1996 *J. Geophys. Res.* **101** 5200

Pécseli H L, Lybekk B, Trulsen J and Eriksson A 1997 *Plasma Phys. Control. Fusion* **39** A227

[166] Strangeway R J *et al* 1998 *Geophys. Res. Lett.* **25** 2065

[167] Ergun R E *et al* 1998 *Geophys. Res. Lett.* **25** 2061

[168] Chang T and Coppi B 1981 *Geophys. Res. Lett.* **8** 1253

[169] Chang T 1993 *Phys. Fluids* B **5** 2646

[170] Bryant D A 1978 *ESA Scientific Publication SP-135* p 52

[171] Bryant D A 1993 *J. Br. Planet. Soc.* **46** 107

[172] Bryant D A, Courtier G M and Bennett G 1972 *Earth's Magnetospheric Processes* ed B M McCormac (Dordrecht: Reidel) p 141

[173] Takeda Y and Yamagiwa K 1991 *Phys. Fluids* B **3** 288

[174] Alfvén H 1987 Keynote address *Double Layers in Astrophysics, NASA Conference Publication No 2469* ed A C Williams and T W Morehead

[175] Montgomery M D 1970 *Particles and Fields in the Magnetosphere* ed B M McCormac (Dordrecht: Reidel) p 95

Montgomery M D, Asbridge J R and Bame S J 1970 *J. Geophys. Res.* **75** 1217

[176] Feldman W C 1985 *Collisionless Shocks in the Heliosphere, Geophysical Monograph 35* (Washington, DC: American Geophysical Union) p 195

[177] Acuna M H, Ousley G W, McEntire R W, Bryant D A and Paschmann G 1985 *IEEE Trans. Geosci. Remote Sensing* **GE-23** 175 and ensuing papers

[178] Bryant D A, Krimigis S M and Haerendel G 1985 *Geosci. Remote Sensing* **GE-23** 177

[179] Shah H M, Hall D S and Chaloner C P 1985 *Geosci. Remote Sensing* **GE-23** 293

[180] Coates A J, Bowles J A, Gowan R A, Hancock B K, Johnstone A D, and Kellock S J 1985 *IEEE Trans. Geosci. Remote Sensing* **GE-23** 287

[181] Southwood D J, Mier-Jedrzejowicz W A C and Russell C T 1985 *Geosci. Remote Sensing* **GE-23** 301

[182] Darbyshire A G, Gershuny E J, Jones S R, Norris A J, Thompson J A, Whitehurst G A, Wilson G A and Woolliscroft L J C *Geosci. Remote Sensing* **GE-23** 311 1985

[183] Feldman W C *et al* 1982 *Phys. Rev. Lett.* **49** 199

[184] Decker R B 1988 *Space Sci. Rev.* **48** 195

[185] Goodrich C C and Scudder J D 1984 *J. Geophys. Res.* **89** 6654

[186] de Hoffman F and Teller E 1950 *Phys. Rev.* **80** 692

[187] Vaisberg D L, Galeev A A, Zastenker G N, Klimov S I, Nozdrachev M N, Sagdeev R Z, Sokolov A Yu and Shapiro V D 1983 *Sov. Phys. JETP* **58** 716

Gurnett D A 1985 *Collisionless Shock Waves in the Heliosphere* ed E Stone and B Tsurutani (Washington, DC: American Geophysical Union)

Fredericks R W, Coroniti F V, Kennel C F and Scarf F L 1970 *Phys. Rev. Lett.* **24** 994

[188] Papadopoulos K 1981 *ESA Scientific Publication SP-161* p 313

[189] Galeev A A *et al* 1986 *Adv. Space Res.* **6** 45

[190] Valenzuela A, Haerendel G, Föppl H, Melzner F, Neuss H, Rieger E, Stöcker J, Bauer O, Höfner H and Loidl 1986 *Nature* **3220** 700

[191] Rees D, Hallinean T J, Stenbaek-Nielsen H C, Mendillo M and Baumgardner J 1986 *Nature* **320** 704

[192] Haerendel G, Valenzuela A, Bauer O H, Ertz M, Foppl H, Kaiser K-H, Lieb W, Loidl J, Melzner F, Merz B, Neuss H, Parigger P, Rieger E, Schönong R, Stöcker J, Wiezorrek E and Molina E 1985 *IEEE Trans. Geosci. Remote Sensing* **GE-23** 253

[193] Krimigis S M *et al* 1986 *J. Geophys. Res.* **91** 1339

[194] Rogers D J, Coates A J, Johnstone A D, Smith M F, Bryant D A, Hall D S and Chaloner C P 1986 *Nature* **320** 712–6

[195] Hall D S, Bryant D A, Chaloner C P, Bingham R and Lepine D R 1986 *J. Geophys. Res.* **91** 1320

[196] Gurnett D A, Ma T Z, Anderson R R, Bauer O H, Haerendel G, Häusler B, Paschmann G, Treumann R A, Koons H C, Holzworth R and Lühr H 1986 *J. Geophys. Res.* **91** 1301
Gurnett D A, Anderson R R, Häusler B, Haerendel G, Bauer O H, Treumann R A, Koons H C, Holzworth R and Lühr H 1985 *Geophys. Res. Lett.* **12** 851

[197] Schwartz S J, Chaloner C P, Christiansen P J, Coates A J, Hall D S, Johnstone A D, Gough M P, Norris A J, Rijnbeek R P, Southwood D J and Woolliscroft L J C 1985 *Nature* **318** 269

[198] Paschmann G 1986 *Minutes of AMPTE UKS-IRM Workshop (Rutherford Appleton Laboratory, November 1986)*

[199] Bryant D A and Riggs S 1989 *Phil. Trans. R. Soc.* A **328** 43

[200] Hapgood M A and Bryant D A 1992 *Planet. Space Sci.* **40** 1431

[201] Ockham's razor of William of Ockham (Occam), Surrey, UK (c. 1285–1349). Familiar as ... 'Entities are not to be multiplied without necessity'; or, from writings, 'It is vain to do with more what can be done with fewer'. 'I have myself found this a most fruitful principle in logical analysis'. (Bertrand Russell)

[202] Fuselier S A, Anderson B J and Onsager T G 1995 *J. Geophys. Res.* **100** 11805

[203] Phan T D, Paschmann G, Baumjohann W, Sckopke N and Lühr H 1994 *J. Geophys. Res.* **99** 121

[204] Heikkila W J 1992 *Proc. Int. Conf. on Substorms (ICS-1), ESA Scientific Publication SP-335* 319

[205] Scholer M 1995 *Physics of the Magnetopause, Geophysical Monograph 90* (Washington, DC: American Geophysical Union) p 235

[206] Lockwood M H and Hapgood H 1997 *Geophys. Res. Lett.* **24** 373

[207] Richmond A D 1987 *The Solar Wind and the Earth* ed S-I Akasofu and Y Kamide (Tokyo: Terra)

[208] Smart D F and Shea M A 1985 *Handbook of Geophysics and the Space Environment* ed A S Jursa (USAF Geophysics Laboratory Publication)

[209] Israel M H 1967 *J. Geophys. Res.* **74** 4701

[210] Smith M J, Bryant D A and T Edwards 1980 *J. Atmos. Terr. Phys.* **42** 167

[211] Blake J B, Kolasinski W A, Fillius R W and Mullen E G 1992 *Geophys. Res. Lett.* **19** 821

[212] Stenzel R L and Gekelman 1984 *Adv. Space Res.* **4** 459

[213] Hones E W Jr, Baker D N, Bame S J, Feldman W C, Gosling J T, McComas D J, Zwickl R D, Slavin J A, Smith E J and Tsurutani B T 1984 *Geophys. Res. Lett.* **11** 5

[214] Burch J L 1987 *The Solar Wind and the Earth* ed S-I Akasofu and Y Kamide (Tokyo: Terra) p 103

[215] Cowley S W H 1982 *Rev. Geophys.* **20** 531

Also Speiser T W 1984 *Adv. Space Res.* **4** 439 and references therein

[216] Kennel C F and Petschek H E 1966 *J. Geophys. Res.* **71** 1

[217] Kellogg P and Van Allen J 1959 *Nature* **183** 1295
Parker E N 1960 *J. Geophys. Res.* **65** 3117
Tverskoy B A 1969 *Rev. Geophys.* **7** 219

[218] Bryant D A, Courtier G M and Bennett G 1973 *Planet. Space Sci.* **21** 165

[219] Heikkila W J and Pellinen R J 1977 *J. Geophys. Res.* **82** 1610

[220] Chapman C S and Watson N W 1993 *J. Geophys. Res.* **98** 165

[221] Parker E N 1957 *J. Geophys. Res.* **62** 509

[222] Sweet P A 1958 *Cosmical Plasmas* ed B Lehnert (Cambridge: Cambridge University Press) p 123
Sweet P A 1969 *Ann. Rev. Astron. Astrophys.* **7** 149

[223] Li X, Roth I, Temerin M, Wygant J R, Hudson M K and Blake J B 1993 *Geophys. Res Lett.* **20** 2423 (theory)

[224] Maran S T 1991 *The Astronomy and Astrophysics Encylopedia* (Cambridge: Cambridge University Press)

[225] Rust D M 1987 *The Solar Wind and the Earth* ed S-I Akasofu and Y Kamide (Tokyo/Singapore: Terra/World Scientific)

[226] Carlowicz M 1997 *EOS Trans. Am. Geophys. Union* **78** 49

[227] McNutt Jr R L, Gold R E, Roelof E C, Zanetti L J, Reynolds E L, Farquhar R W, Gurnett D A and Kurth W S 1997 *J. Brit. Interplanet. Soc.* **50** 463

[228] Pilipp W G, Miggenrieder H, Montgomery M D, Mühlhäuser K-H, Rosenbauer H and Schwenn R 1987 *J. Geophys. Res.* **92** 1075

[229] Scudder J D 1994 *Astrophys. J.* **427** 446
Scudder J D and Olbert S 1979 *J. Geophys. Res.* **84** 2755
Scudder J D and Olbert S 1979 *J. Geophys. Res.* **84** 6603

[230] Bryant D A, Cline T L, Desai U D and McDonald F B J 1962 *Geophys. Res.* **67** 4983
Bryant D A, Cline T L, Desai U D and McDonald F B 1965 *Astrophys. J.* **141** 478

[231] Hamilton D C 1977 *J. Geophys. Res.* **82** 2157

[232] Cargill P 1996 *EOS Trans. Am. Geophys. Union* **77** p 353
Simnet G M 1996 *EOS Trans. Am. Geophys. Union* **77** p 355
Emslie A G 1996 *EOS Trans. Am. Geophys. Union* **77** p 355

[233] Bingham R, Su J J, Dawson J M, McClements K G and Spicer D S 1993 *Astrophys. J.* (May)

[234] Fermi E 1950 *Nuclear Physics* rev. edn (Chicago, IL: University of Chicago Press)

[235] Clay R and Dawson B 1997 *Cosmic Bullets* (St Leonards, NSW: Allen and Unwin)

[236] Linsley J 1980 *Origin of Cosmic Rays* ed G Setti, G Spada and A W Wolfendale (Dordrecht: Reidel) p 53

[237] Schever P A G 1984 *Adv. Space Res.* **4** 337

[238] See 1992 *Phys. Rev.* D (part 11) **45** 111.44

[239] Nishimura J 1980 *Origin of Cosmic Rays* ed G Setti, G Spada and A W Wolfendale (Dordrecht: Reidel) p 75
Cowsik R 1980 *Origin of Cosmic Rays* ed G Setti, G Spada and A W Wolfendale (Dordrecht: Reidel) p 93

[240] Fermi E 1954 *Astrophys. J.* **119** 1

[241] Achterberg A 1984 *Adv. Space. Res.* **4** 193

Questions and exercises

Questions

1. What was the accelerator of Isaac Newton's falling apple?

2. Electrons are accelerated within electron multipliers such as photomultipliers, channel electron multipliers and channel plates. Identify the non-conservative element(s) of the process.

3. Particles radiate from an object into free space. What happens to velocity-space density as distance from the source increases?

4. A (nearly) parallel beam of photons is brought to a (near) focus by a lens. Does this increase the phase-space density?

5. Low-rigidity cosmic rays unable to penetrate through the geomagnetic field to reach the equatorial atmosphere can do so in the polar regions. How would you expect the intensity as measured with a highly directional detector to vary when measured from a satellite en route from the pole to the equator?

6. Within the magnetopause boundary layer we observe velocity-space densities intermediate between those of the magnetosheath and the magnetosphere. Is this consistent with Liouville's theorem? Is the fall of velocity-space density with time and distance from the origin, for example following the release of solar-flare particles, consistent with Liouville's theorem when the scattering is caused by purely dynamical processes?

7. Devise a statement citing Liouville's theorem to prove that two streams of particles following dynamical trajectories cannot merge.

8. Two flat discs slide towards each other along a common line towards each other and, on colliding, exchange some energy. Which is the energy donor?

9. Do solar wind ions cool down as the result of expanding into an ever-increasing volume?

10. How would figure 1.32 appear if the electrons had to surmount the potential barrier before being accelerated away?

11. Sketch an equivalent of figure 1.33 for a situation in which electrons were accelerated into a potential well before having to climb out of it.

12. It has been suggested more than once in defence of the potential-difference theory of auroral electron acceleration that the equipotentials close at the Sun. Is this a valid defence?

13. How would figure 1.47 differ if it were plotted in terms of intensity rather than velocity-space density?

14. Something is missing from figure 1.11. Please point out what it is, and confirm that if it were included, it would not alter the conclusions drawn.

15. What is the significance for velocity-space density of the colour-coded peak intensities remaining almost at a constant colour throughout each of the auroral arc spectrograms?

16. Describe the appearance of a figure equivalent to figure 1.46 with the ordinate being intensity rather than velocity-space density. Describe separately cases where the low-energy particles had (a) Maxwellian, (b) kappa or (c) power-law distributions.

17. If a process in which cosmic rays undergo random gains and losses in collisions with wandering magnetic fields were to continue indefinitely, would it lead to equipartition of energy between particles and magnetic clouds?

18. How much energy could a wavepacket of maximum potential excursion ϕ impart to an electron in a single interaction under optimum resonance conditions? If $\Delta = e\Phi$, is it (a) less than Δ, (b) equal to Δ, (c) more than Δ or (d) unlimited?

19. An electron stream passes through a medium where exposure to wavepackets is strong enough for the stream to attain an equilibrium energy distribution. The phase and group velocities of the component waves are equal and are the same for all wavepackets. What, according to the wave theory, are the requirements for an originally monotonically falling energy distribution to develop a peak?

20. Co-rotation of plasma with a magnetised planet causes magnetic lines of force becoming electrostatic equipotentials. The strength of the consequent electric field appears, then, to be determined on the one hand by the spacing between field lines, and on the other hand by the cross product of the local co-rotation speed and magnetic field. Are these consistent?

Exercises

E1. Convince yourself that the generation of the peak in figure 3.31(*e*) is accompanied by an increase in entropy of the electron velocity distribution.

E2. Make a transparency of the set of parallel lines representing a lower-hybrid wave in figure 3.27, to demonstrate that the parallel velocity of lower-hybrid waves greatly exceeds the perpendicular phase velocity.

E3. Make a transparency of the longer wavelength 'waves' of figure 3.32, to demonstrate that for waves whose phase velocity increases with wavelength (as is the case for lower-hybrid waves), the velocity of beats—the group velocity—is lower than the phase velocity of either of its component waves. For good measure, confirm that for waves whose phase velocities decrease with wavelength, the reverse is the case.

E4. Explore with the aid of the QBasic program in Appendix 1 how, when particle numbers and total energy are conserved, increasing entropy drives a distribution into the form of a Maxwellian, which then remains stable. Demonstrate, for example, how the speed of evolution increases with increased 'exposure' and with larger individual energy changes.

Note that with all of these Monte-Carlo experiments it will probably be found convenient to begin with a small number of particles, and to increase this number (and with it the total number of operations) in refinements to improve statistical accuracy only as the desired conditions are approached.

E5. Explore with the QBasic program in Appendix 2 how random encounters between particles lead, on exchange of all or part of their energies, to increasing entropy and to the evolution of Maxwellians. Demonstrate that evolution is faster when energy exchanges are greater, and investigate the role of statistical fluctuations in allowing further changes to occur even after a near-Maxwellian has been attained.

E6. Investigate, using the QBasic program in Appendix 2, how particle velocity distributions can become modified by resonant interaction with waves. Show, in particular, that the naturally increasing entropy allows peaks to form in the distribution, and that peak formation is encouraged by steeper initial particle distributions and by larger amplitude and narrow-band waves. Show how broader band waves and long 'exposures' tend to promote general heating.

Symbols, abbreviations and units

Symbols

a	acceleration
A	area
B	magnetic flux density or 'field strength'
c	speed of light in vacuo ($= 3.00 \times 10^8$ m s^{-1})
c	count rate
C	compression ratio
d	dimension
e	elementary charge ($=1.60 \times 10^{-19}$ C)
E	kinetic energy
E_κ	characteristic energy of kappa distribution
E_M	characteristic energy of Maxwellian distribution
E_0	energy equivalent of rest mass ($= 5.11 \times 10^5$ eV for electron)
$\hat{E}$	most probable energy
$\langle E \rangle$	mean kinetic energy
$\mathcal{E}$	electric field
f	velocity-space density
F	phase-space density
$\mathcal{F}$	force
G	gain factor
j	differential intensity
k	Boltzmann's constant ($= 1.38 \times 10^{-23}$ J K^{-1})
K	wavenumber
l, L	length
m	mass
m_e (or m)	electron mass ($= 9.11 \times 10^{-31}$ kg)
M	ion mass
$\mathcal{M}$	magnetic moment
n	number density

298

N	number
N_D	number in Debye sphere
N_0	expected number
p	momentum
p_x, p_y, p_z	coordinates of momentum space
P	power
$\mathcal{P}$	power density
r, R	radius or radial distance
r_c	gyro radius
r_{cv}	radius of curvature
R_E	a distance equal to the Earth's mean radius ($= 6370$ km)
S	entropy
t	time
T	absolute temperature
T_{eff}	effective temperature
$\mathcal{T}$	volume in real and velocity space combined
u	velocity, speed of specified object
U	velocity
v	velocity, speed
v_g	group velocity
v_x, v_y, v_z	coordinates of velocity space
v_ϕ	phase velocity
V	volume in real space
$\mathcal{V}$	'volume' in velocity space
$\mathcal{V}$	'volume' in momentum space
w	velocity
x, y, z	coordinates of real space
X, Y, Z	Geocentric, Solar, Ecliptic (GSE) coordinates
Z	number of units of electronic charge
α	pitch angle
β	velocity as fraction of velocity of light in vacuo
γ	index of power-law spectrum
Γ	relativistic (Lorentz) factor
Δ	a specified quantity of energy
ε	energy as a fraction of rest energy
ε_0	permittivity of free space ($= 8.855 \times 10^{-12}$ F m^{-1})
η	thickness
θ	angle
κ	parameter of kappa distribution
K	energy associated with kappa distribution ($= \kappa E_\kappa$)
λ	wavelength, mean free path
λ_D	Debye length in a Maxwellian plasma
λ_κ	Debye length in a kappa distribution plasma

Λ	latitude
μ	magnetic moment of gyrating particle
μ_0	permeability of free space ($= 4\pi \times 10^{-7}$ H m^{-1})
ν	frequency
ξ	azimuthal angle
Ξ	energy density
π	ratio of circumference of circle to diameter ($= 3.142\ldots$)
ρ	mass density
P	magnetic rigidity
τ	period
τ_c	gyro period
τ_p	plasma oscillation period
$\mathcal{T}$	volume in phase space
ϕ	magnetic flux
Φ	electrostatic potential
ψ	probability
ω	angular frequency
Ω	solid angle
d$*$	vanishingly small increment of quantity $*$
$\delta*$	range of quantity $*$
$\Delta*$	change in quantity $*$
$\sum *$	sum of all quantities of type $*$
e	base of natural logarithms ($= 2.718\ldots$)
grad	spatial gradient of
$\parallel$	parallel
$\perp$	perpendicular
#	anti-parallel
$\otimes$	arrow into paper
$\odot$	arrow out of paper
!	factorial

Subscripts and abbreviations

A	Alfvén
AMPTE	Active Magnetospheric Particle Tracer Explorers
BL	boundary layer
BS	bow shock
c	gyro (or cyclotron), also centre of mass
CCE	Charge Composition Explorer of the AMPTE mission
CLRC	Central Laboratory of the Research Councils

CME	Coronal Mass Ejection
coll	collisions
CR	cosmic ray
CRA	cosmic ray albedo
cv	curvature
D	Debye
DE	Dynamics Explorers
DL	depletion layer
e	electron
FAST	Fast Auroral Snapshot Satellite
g	group
GSE	Geocentric, Solar, Ecliptic
i	ion
I	ionosphere
IMP	Interplanetary Monitoring Platform
IRM	Ion Release Module of the AMPTE mission
ISEE	International Sun-Earth Explorer
J	Jupiter
lh	lower-hybrid
LLBL	low-latitude boundary layer
M	Maxwellian
max	maximum
MP	magnetopause
MSH	magnetosheath
MSP	magnetosphere
NR	non-relativistic
OGO	Orbiting Geophysical Observatory
p	plasma
PS	plasma sheet
PSBL	plasma-sheet boundary layer
R	random
RB	radiation belt
rel	relative
rms	root mean square
S	systematic
SW	solar wind
SWE	solar-wind event
TL	tail lobe
UKS	United Kingdom Satellite of the AMPTE mission
UR	ultra-relativistic
UT	universal time
ϕ	phase

Units

AU	astromical unit $(= 1.50 \times 10^{11}$ m)
eV	electron volt $(= 1.60 \times 10^{-19}$ J)
J	joule
K	kelvin
kg	kilogram
m	metre
N	newton
rad	radian
s	second
sr	steradian
T	tesla
V	volt

Prefixes

z	zepto	10^{-21}
n	nano	10^{-9}
μ	micro	10^{-6}
m	milli	10^{-3}
k	kilo	10^{3}
M	mega	10^{6}

Answers to questions

1. The Sun. Solar radiation raised the apple's material against the force of gravity to be absorbed by the tree before the work done was repaid by gravity when the apple fell. A similar description applies to waterfalls.

2. The essential non-conservative element is secondary emission. If the detector is powered by a battery, chemical reactions within the battery form another element.

3. Since the volume of velocity space decreases in proportion to the solid angle subtended by the source, and the volume of real space increases with the square of radial distance, and these are in the same proportion, velocity-space density remains constant—as Liouville's theorem advises immediately.

4. The wider range of directions at the focus increases the volume of momentum space in the same proportion that focusing reduces the volume occupied in real space, so the phase-space density remains constant—again, as Liouville's theorem advises.

5. The intensity would remain at the level observed over the pole until a cut-off latitude was reached at which it would fall to zero. This follows from Liouville's theorem which holds that if there is a trajectory at all, the velocity-space density will not be changed from that in outer space. Since there is no acceleration, intensities will likewise remain unchanged. Intermediate intensities will be registered close to the cut-off latitude where the detector's field of view is only partially filled with allowed trajectories.

6. Yes, they are consistent. Although random scattering will not modify velocity-space density along trajectories, the trajectories may not fill the whole of phase space, thus causing densities observed over a finite volume of phase space to differ from those of the source, or sources.

7. Since, by Liouville's theorem, velocity-space densities remain constant along both dynamical trajectories, and merging would require velocity-space densities to increase, merging cannot take place. (... or similar)

8. The energy donor is the one with the greater initial magnitude of momentum in the observer's frame of reference.

9. No work is done in unhindered expansion, so no change of temperature results. In practice, dynamic scattering from irregularities in heliospheric space and other processes can cause heating [241].

10. It would be the same as the original, except for a cut-off at low energies of those electrons unable to surmount the barrier. The cut-off would be at the same energy as the peak in the figure.

11. Since energy gained on the way in would be lost on the way out, the emerging distribution would be the same as the initial one.

12. Since the inability of closed contours to provide acceleration is a topological fact, it is immaterial where the contours close—the net effect is always zero.

13. Since f and j at a given energy bear a constant relation to each another, the traces would unchanged in shape. They would, though, on the same logarithmic scale, be more closely spaced when plotted in terms of intensity.

14. What is missing is the return magnetic flux. Since lines of force are continuous, the net magnetic flux through any closed surface or infinite plane is zero, so the increase in into-the-paper flux shown will be balanced by an equal and opposite increase out of the paper. If the return is at a remote location (as implied in the figure) its consequent electric fields will only negligibly reduce the field in the region shown. However, if the return is nearby they could significantly modify the picture. The conclusion that there is an energy-changing electric field—would, however, be nullified only in the limit of co-location, in which case there would be no changing magnetic field.

15. The approximately constant peak intensity reveals immediately that velocity-space density at the peak falls as the energy at which it occurs rises.

16. In all cases the peak would be more pronounced, i.e. the peak-to-valley ratio would be greater. If the central core were Maxwellian or followed a kappa distribution, intensities would fall to zero at zero speed, thus giving the whole figure the appearance of a volcano. A power-law distribution at the core would be qualitatively unchanged for negative exponents less than unity. There would be a flat plateau for a negative exponent of unity, and the distribution would rise indefinitely for a negative exponent greater than unity.

17. No, the trend is towards the maximum entropy that can be attained in the steady state in which fresh particles are continuously introduced. There is no *a priori* reason why this should correspond to a state of equipartition. Fermi pointed out, in fact, that equipartition would lead to 'unbelievable' cosmic-ray energies. It may be worth contemplating, though, with the aid of figure 10.5, that if the controlling product $N_0\beta^2$ were to exceed unity, the negative power law exponent would—for the processes discussed here—fall to a value near 4, when the high-energy content of the distribution would tend to infinity. The number content would diverge for an exponent of 3.

18. The answer is (c), more than Δ. The method employed in figure 1.4 can be employed to show that very low energy electrons can gain 4Δ, and a similar exercise using momentum instead of velocity will show that there is an ultra-relativistic limit of $2\sqrt{2E_0\Delta}$. The general expression for the increase in energy under optimum resonance conditions is

$$\sqrt{E_0^2 + \left[\sqrt{2E_0E + E^2} + 2\sqrt{2E_0\Delta + \Delta^2}\right]^2} - (E + E_0).$$

19. The wavepacket amplitude needs to be large enough for the mean velocity-space density at the limits of resonance (or two other complementary points) to exceed the mean velocity-space density at two complementary points closer to the wavepacket velocity. This condition is met if the energy distribution in the region has a negative power-law slope of greater than unity.

20. Consider a planet with a cylindrically symmetric, but not necessarily dipolar, magnetic field and let the rotation and magnetic axes coincide. Take an elemental magnetic flux tube of content ϕ. Let it have at a distance R from the axis of rotation, a longitudinal extent $R\xi$, where ξ is a small azimuthal angle. Its sides in a direction normal to this will be $\phi/BR\xi$, where B is the magnetic field strength. The (perpendicular-to-B) electric field required to counter the magnetic part of the Lorentz force is of magnitude ωBR, where ω is the angular rate of rotation of the planet. The potential difference along the sides is just this field multiplied by the length of the sides or $\phi\omega/\xi$. Since this is independent of both R and B, it holds anywhere along the flux tube, signifying that velocity, magnetic field, electric field and line-of-force spacing are automatically consistent with lines of force being electrostatic equipotentials.

It is clear that, while a slight tilt of the magnetic axis with respect to the rotation axis (as in the case of the Earth's $11°$) might still approximate to this situation, gross departures from symmetry would lead to reversals of directions of motion within flux tube and, consequently, gross departures from simple co-rotation.

Index